U0054091

應用英文寫作系列之七

Effective Management Communication
管理英文

By Ted Knoy
柯泰德

Ted Knoy is also the author of the following books in the Chinese Technical Writers' Series（科技英文寫作系列叢書）and the Chinese Professional Writers' Series（應用英文寫作系列叢書）：

An English Style Approach for Chinese Technical Writers
《精通科技論文寫作》

English Oral Presentations for Chinese Technical Writers
《做好英文會議簡報》

A Correspondence Manual for Chinese Writers
《英文信函參考手冊》

An Editing Workbook for Chinese Technical Writers
《科技英文編修訓練手冊》

Advanced Copyediting Practice for Chinese Technical Writers
《科技英文編修訓練手冊進階篇》

Writing Effective Study Plans
《有效撰寫讀書計畫》

Writing Effective Work Proposals
《有效撰寫英文工作提案》

Writing Effective Employment Application Statements
《有效撰寫求職英文自傳》

Writing Effective Career Statements
《有效撰寫英文職涯經歷》

Effectively Communication Online
《有效撰寫英文電子郵件》

Writing Effective Marketing Promotional Materials
《有效撰寫行銷英文》

本書獻給我的岳父　黃國榮 先生

Foreword

Professional writing is essential to the international recognition of Taiwan's commercial and technological achievements. The Chinese Professional Writers' Series seeks to provide a sound English writing curriculum and, on a more practical level, to provide Chinese speaking professionals with valuable reference guides. The series supports professional writers in the following areas:

Writing style

The books seek to transform old ways of writing into a more active and direct writing style that better conveys an author's main ideas.

Structure

The series addresses the organization and content of reports and other common forms of writing.

Quality

Inevitably, writers prepare reports to meet the expectations of editors and referees/reviewers, as well as to satisfy the requirements of journals. The books in this series are prepared with these specific needs in mind.

Effective Management Communication is the seventh book in The Chinese Professional Writers' Series.

"Effective Management Communication"(《管理英文》)爲「應用英文寫作系列」(The Chinese Professional Writers' Series)之第七本書，本書中練習題部分主要是幫助國人糾正常犯寫作格式上錯誤，由反覆練習中，進而熟能生巧提升有關個人職涯描述的英文寫作能力。

「應用英文寫作系列」將針對以下內容逐步協助國人解決在英文寫作上所遭遇之各項問題：

A.寫作形式：把往昔通常習於抄襲的寫作方法轉換成更積極主動的寫作方式，俾使讀者所欲表達的主題意念更加清楚。更進一步糾正國人口語寫作習慣。

B.方法形式：指出國內寫作者從事英文寫作或英文翻譯時常遇到的文法問題。

C.內容結構：將寫作的內容以下面的方式結構化：目標、一般動機、個人動機。並瞭解不同的目的和動機可以影響報告的結構，由此，獲得最適當的報告內容。

D.內容品質：以編輯、審查委員的要求來寫作此一系列之書籍，以滿足讀者的英文要求。

Introduction

This handbook orients managers on the essentials of written communication for managers in the workplace. How to write effective workplace memos is described, including recommendation reports, persuasive reports and informal technical reports. How to effectively communicate online is then introduced by outlining common forms of correspondence, including technical training, information exchanges, overseas technical visits, speaker and consultant invitations and information requests. Next, how to write an effective career statement for employment is described, including how to express interest in a profession, describe the importance of a particular topic within one's profession, describe participation in a related project, describe one's academic background, introduce one's research and professional experiences, as well as describe one's civic and extracurricular activities. Additionally, how to write an effective work proposal is introduced, including a description of the project background and plan, problem statement, hypothesis statement and abstract. Moreover, how to write an effective academic or professional training proposal is introduced. Also, how to deliver an effective oral presentation is introduced, including helpful phrases for the introduction, body and conclusion. Different types of oral presentations are also introduced, including the forecasting of market trends, description of product or service development, project description for developing a product or service, technology introduction and introduction of an industry. Additionally, how to edit a technical manuscript for conciseness and clarity is introduced, as well as for common Chinese-English colloquial habits in writing. The final three units describe how to write employment application letters, training applications letters and employment recommendation letters.

簡　介

本手冊主要訓練管理人才（管理師）撰寫符合工作場合需要的書面英文。前三單元，有效撰寫管理英文備忘錄，包括建議性報告、說服力的展現及非正式實用管理技術報告。第四單元，英文書信溝通包括：科技訓練請求信函、資訊交流信函、海外技術訪問、邀請演講者信函及資訊請求信函，來建構如何有效的和海外專業人士交流英文。第五單元，工作經歷撰寫主要強調以下部分：表達工作相關興趣、工作主題的專業性、所參與方案裡專業興趣的表現、學歷背景、研究及工作經驗及與求職相關的課外活動。第六單元，有效撰寫英文工作提案包括：方案背景描述、問題描述、假設描述及摘要撰寫。第七單元，管理師學術及專業訓練英文申請撰寫，主要探討管理師如何撰寫學術及專業訓練英文申請。第八單元，如何有效進行管理英文口語簡報強調有關簡介、本體及結論部分的有用詞彙，同時也介紹不同形式的口語簡報，包括：預測市場趨勢、產品或服務研發、專案描述、技術介紹及工業介紹。第九單元，如何以精確寫作及明白寫作原則編修科技英文寫作手稿，以及中英口語寫作陋習糾正。第十單元，求職申請信函，包括：經由廣告、朋友及其他相關訊息得知的工作申請，對未來工作的個人學術及相關專業特質總結，讚揚要申請工作的機構及說明其對個人工作目標所產生的利益，總結主要論點及邀請讀者給予面試機會。第十一單元，專業訓練申請信函，包括：專業訓練申請，某人學歷及工作經驗概述，讚揚提供專業訓練的機構，專業訓練細節解釋及結語。第十二單元，求職推薦信函，包括：簡介、推薦人的資格、被推薦人跟求職有關的個人特質及信函結尾。

Table of Contents

Table of Contents

Table of Contents

Unit One

Writing Effective Workplace Memos: Recommendation Reports

有效撰寫管理英文備忘錄——調查性與建議性報告

簡 介

工作備忘錄是台灣管理師和國外相對合作對象溝通的必需文件，本單元描述兩種常見的工作備忘錄：調查性報告及建議性報告。調查性報告分析必須解決的管理問題，建議性報告則概述目標、方法論及預期的結果。

一、調查性報告：告知讀者特定管理問題的背景及合乎邏輯的需求回應，主要寫作論點：

A.簡介：您的機構跟這個主題的相關程度有多大？

範例 i.

During our most recent board meeting, participants expressed concern over the current trend in financial institutions to underrate a company's value owing to the lack of consideration of intellectual capital assets, which dominate in enterprises such as online gaming companies.

範例 ii.

At a recent meeting, members discussed the increasing incidence of outstanding debts and a higher risk of defaulted loans owing to substandard lending practices that have significantly undermined the creditworthiness of issuing banks.

範例 iii.

In our working group meeting last week, we discussed our hospital's difficulty in retaining current patients and reducing the turnover rate of those going to other medical centers for treatment.

範例 iv.

Hospital administrators expressed concern recently over how to select the most qualified vendor for handling medical waste produced by our facility.

範例 v.

At a recent meeting, members discussed ways in which to enhance the performance of hospitals so that administrators can revise directions in management based on analysis methods that accurately reflect the efficiency of healthcare services provided.

範例 vi.

Our hospital administrators expressed concern recently over how to evaluate outsourced nursing care attendants when selecting the most productive outsourcing agencies.

B.目前的管理界情勢：管理問題後的工業背景？

範例 i.

With the global economy moving into the information age, knowledge is the most fundamental intangible capital asset and will increasingly dominate efforts to create a competitive edge and generate wealth. Although tangible assets such as property, facilities and equipment continue to play a vital role in manufacturing products and providing services, their relative importance has declined with the increasing importance of intellectual capital. Intellectual capital takes the form of inventions, ideas, general know-how, design approaches, computer programs, processes and publications.

範例 ii.

The increasing popularity of credit cards in Taiwan has led to stringent competition domestically. Banks encourage prospective customers to hold multiple credit cards by relaxing approval and credit reference procedures.

範例 iii.

The extremely competitive medical market sector in Taiwan and budget deficits incurred from the island’s National Health Insurance scheme have led to the implementation of a Global Budget System. Whereas hospitals heavily prioritize attracting new patients, the patient turnover rate has received increasing attention. Thus, hospitals must concentrate their operational efforts on retaining current patients and reducing the turnover rate of those going to other medical centers for treatment.

範例 iv.

Taiwanese hospitals must effectively cope not only with severe budgetary constraints in the national health insurance scheme, but also with stringent mandates on medical waste treatment from the Environmental Protection Administration. While hospitals dispose of large amounts of medical waste to ensure sanitation and personal hygiene, doing so inefficiently creates potential environmental hazards and increases operational expenses. The environmental protection sector is just beginning to address harmful waste that the waste treatment sector is emitting, largely owing to the lack of effective management experience in closely monitoring daily operations.

範例 v.

Given rapid changes in Taiwan's healthcare environment, limited medical resources make efficient distribution essential, which is of priority concern among health organizations. State level administrators should analyze the productivity of each hospital to determine whether resources are utilized effectively.

範例 vi.

A rapidly growing elderly population poses a major challenge for long-term care management, requiring immediate solutions given changing family structures and the frequency of chronic illnesses. Employees in hospital subsidized respiratory care centers, respiratory care wards and nursing homes are nearly all nursing care attendants. Belonging to a sub-specialized field of a medical department, nursing care attendants do not hold specialized license certification nor have a certain academic level. These workers simply require basic healthcare training skills and knowledge of hospital or governmental infrastructure to perform their tasks efficiently. Playing an important role in Taiwanese society, nursing care attendants accompany disabled individuals and provide assistance in daily activities such as taking a bath, taking meals, monitoring urinary or stool specimens, changing the posture of incapacitated stroke victims and providing general comfort.

C.管理問題：您的機構要解決什麼樣的問題？

範例 i.

Although capable of determining the value of tangible assets that can be quantified in a company, conventional accounting methods do not include intellectual capital, subsequently inducing an underrating of a company's value. Therefore understanding exactly how an organization assesses its intellectual competency is of priority concern. Additionally, scientific and technological advances have heightened the importance of intellectual capital, as evidenced by the numerous companies that rely almost completely on intellectual assets for generating revenue.

範例 ii.

However, such practices have significantly undermined the creditworthiness of issuing banks, leading to increased outstanding debts and a higher risk of defaulted loans. In this respect, quantitatively measuring the varying degrees of risk among potential credit card customers to identify both varying degrees of risk and reasons for credit card debt has seldom been performed in previous studies.

範例 iii.

Of the relatively few studies that have addressed the medical sector in this area, conventional approaches estimate the number of patients on a daily basis without differentiating between new and return ones.

範例 iv.

The scope of services that medical waste disposal firms provide does not always comply with the needs of individual hospitals. According to the Environmental Protection Administration, although 456 medical waste disposal firms were operating by the end of 2000, their small-scale size limited the scope of services in handling certain wastes. Therefore, the ability to efficiently handle medical waste depends on the ability of hospitals to adopt sound waste management practices and coordinate their efforts with waste disposal firms. However, Taiwanese hospitals lack an objective criterion to select the most appropriate waste disposal firm and evaluate its performance, instead relying on their own subjective judgment and previous experiences.

範例 v.

In many countries, the public sector rather than non-profit organizations provides most commercial and social services, making their contribution extremely difficult to determine in terms of conventional aggregative performance indicators such as return on investment (ROI), residual income (RI) and profitability. All hospitals utilize resources to provide many services, i.e., a measure of the output for healthcare organizations. Thus, assessing the efficiency of how well input is used to produce these services is an important measure of performance.

範例 vi.

Despite their contributions, most nursing care attendants lack a uniform management approach to ensure the quality of service given that these workers have no restrictions on age, education or experience; they are reguired to be mentally and physically sound. Despite the abundance of outsourcing agencies for nursing care attendants, the lack of standardized training makes it impossible to control the quality of services provided island wide. Moreover, changes in Taiwan's National Health Insurance scheme have led to an emphasis on controlling personnel costs while maintaining highly quality services, further contributing to the outsourcing of nursing care attendants. Generally, relatives or the patient directly employs nursing care attendants without adequate evaluative criteria to select the most appropriate care provider. Additionally, outsourcing firms also lack objective criteria in selecting nursing care attendants, leading to widespread customer dissatisfaction and increasing management difficulty.

D.問題的嚴重性：這個問題影響到您機構的程度有多深？

範例 i

For instance, on-line gaming companies emphasize the ownership of intangible capital rather than tangible assets. Online gaming ranks at the top of the gaming industry, with generated revenues of 1 billion US Dollars in 1999 skyrocketing to 2 billion US dollars in 2002. According to the International Data Corporation (2003), in 2002, the on-line gaming market scale was 533,000,000 in the Asian Pacific region, with South Korea and Taiwan leading the way by accounting for 54% and 26%, respectively.

範例 ii.

To illustrate the severity of this problem, Taiwan has an average of 2.8

cardholders from the currently 6,321 credit cards available domestically, with credit card debt reaching 46%.

範例 iii.

Incapable of determining patient turnover rates precisely, our hospital could expend considerable costs in attracting new patients in the face of shrinking subsidized revenues from the National Health Insurance scheme, ultimately lowering our competitiveness.

範例 iv.

According to Department of Health statistics, with approximately 17,500 hospital clinics operating domestically with a bed capacity of roughly 120,000 beds, nearly 300 metric tons of medical waste are produced daily; among which, 15% is infectious, with its amount increasing as well.

範例 v.

Of the 14,474 public and private medical institutions operating in Taiwan as of the end of 1992, public and private medical care institutions comprised ninety seven hospitals as well as 479 clinics and 728 hospitals as well as 13,170 clinics, respectively.

範例 vi.

Given the twenty two hospital subsidized nursing homes and eighteen respiratory care wards currently operating in Taiwan, the importance of nursing care attendants is obvious.

E.問題的牽涉性：這個沒有解決的問題會如何影響到您的機構及管理業界其他人？

範例 i

The inability to determine the value of companies without incorporating intellectual capital will not only lead to an underrating of their value, but also make it impossible to understand how intellectual capital impacts the on-line gaming market.

範例 ii.

The inability to identify both varying degrees of risk and reasons for credit card debt will further exacerbate the bank lending crisis in Taiwan, ultimately threatening the survival of credit cooperatives owing to the inability to effectively control spiraling outstanding debts.

範例 iii.

The inability to effectively address patient turnover rate would invalidate the statistical approaches used to determine the samples in pertinent studies.

範例 iv.

Given the trend of increasing medical waste, the inability of hospitals to adopt an objective means of evaluating a waste disposal firm and its performance may lead to an inappropriate selection and higher operational costs, necessitating the development of an effective means of evaluating the performance of waste disposal firms objectively, thus reducing overhead costs and enhancing medical waste management.

範例 v.

Given the substantial amount of resources that governments invest in healthcare, maximizing the efficiency of public hospitals is of priority concern. The inability to do so makes it nearly impossible to allocate medical resources to hospitals effectively from a governmental level.

範例 vi.

The inability to manage effectively the quality of service that nursing care attendants provide will lead to a further decline in hospital revenues, eventually leading to a reduction in personnel and community services.

F.管理需求：要如何進行來解決這個管理問題？

範例 i.

Therefore, a novel index based on the analysis hierarchy process (AHP) must be developed to determine the intellectual capital value of on-line gaming companies.

範例 ii.

Therefore, a novel evaluation method must be developed to identify target credit card customers in which the characteristics associated with the lifestyles of such customers and factors that contribute to easily incurred debts are incorporated.

範例 iii.

Therefore, a novel prediction model must be developed to identify the turnover rate of customers in the medical sector by incorporating a novel set of management strategies that emphasize customer retention within the medical sector to strengthen our hospital's competitiveness.

範例 iv.

Therefore, as most hospitals contract waste disposal firms to handle their waste, administrators need a reliable method to select the most appropriate contracted firm in order to closely monitor and reduce expenditures associated with this process.

範例 v.

Therefore, an analysis must be made not only of the efficiency of healthcare services that public hospitals offer using data envelopment analysis (DEA), but also of those factors to enhance the performance of hospitals so that administrators can revise directions in management accordingly.

範例 vi.

Therefore, a selection model based on fuzzy theory and the AHP method must be developed, capable of providing an objective means of evaluating the quality of nursing care attendants, ultimately enhancing the quality of services, increasing customer satisfaction and lowering personnel costs.

二、建議性報告告知讀者管理目標、方法論及預期的結果，以求解決特定的管理問題。

A.問題的總論：該解決的管理問題是什麼？

範例 i.

In our recent working group, members raised the concern that, with respect to differentiated marketing practices in Taiwan, conventional methods of ranking customers are normally based on the bank account balance for each accounting period. This basis alone does not provide a complete customer profile, and seldom incorporates strategies that analyze the commercial transaction data of customers. Insufficient information of unique customer characteristics can obviously not provide specialized services for individuals. Thus, the inability to interact compatibly with customers will cause our company to lose its focus on product development and promotional strategies.

範例 ii.

Taiwan's growing elderly population has increased the demand for long-term healthcare facilities and services. While forecasting the medical market trends (both in supply and demand) is a fundamental aspect of feasibility analysis, both governmental policymakers of social welfare trends and commercial investors heavily rely on forecasting reports to remain abreast of regulations governing health finance policies and to develop inventory projects. Whereas the modeling method has seldom been adopted to forecast market trends in long-term care in Taiwan, most studies focus mainly on various consumer indexes used to forecast the market supply and demand yet neglect those market and natural factors that are related to the long-term healthcare sector.

範例 iii.

Purchasing on credit is increasingly common in Taiwan, as evidenced by the rise in customer loans granted by local banks in recent years. However, when processing the amount of loan applications, the banking officer frequently scores a customer's credit rating based on a manual standard that lacks objectivity and expends a considerable amount of human resources.

範例 iv.

According to the World Health Organization, Taiwan became an advanced aging society as of the end of 1993, with over 7% of its population 65 years or older.

The aging trend in Taiwan explains the growing market demand for long term care residential communities, in which middle to senior aged individuals live independently in houses or apartment units designed for their special needs upon retirement. Many industries have already heavily invested in this emerging growth sector. However, while evaluating the success of certain management practices in satisfying consumer demand in this growing market sector, long term care studies have seldom attempted to identify success factors in managing senior citizen residential communities.

範例 v.

Since its establishment in 1995, Taiwan's National Health Insurance scheme has strived to provide medical coverage for all island residents under the auspices of the National Health Insurance Bureau. Despite the high quality medical care provided, hospitals often have difficulty in understanding customer needs. More specifically, conventional statistical methods such as AHP and TOPSIS cannot analyze how patients select a hospital for treatment, making it extremely difficult to devise effective business management strategies.

範例 vi.

Unforeseeable circumstances in the constantly fluctuating business climate necessitate that enterprises adopt effective inventory management practices to strengthen their competitive edge. Inventory stock is often viewed as a somewhat static resource of economic value, with the quality of its management directly impacting company operations. Therefore, an effective stock inventory system is essential in the supply chain of commercial activities. However, while focusing on specific stock inventory systems to achieve the most appropriate design, previous studies have seldom addressed the supplying chain strategy for stock inventory of multiple products and many suppliers. Either simulation or statistical approaches are adopted to establish inventory stock management policies. Simulation requires much time, making it inefficient with questionable results. Statistical approaches can only offer guidelines aimed at simplifying either the inventory stock administrative system or a complex system intended to establish approximate mathematical patterns; such approaches are often difficult to interpret.

B.建議：您的機構對解決問題有何提議？

範例 i.

Therefore, we recommend developing a customer ranking model by analyzing the dynamic purchasing behavior of customers and identifying the potential ones who can generate bank revenues. Those behavioral results can be used to devise diverse promotional strategies or customize products or services according to consumer needs, thus achieving market differentiation and effective management of customer relations.

範例 ii.

Therefore, we recommend developing two forecasting models to estimate the market demand of the elderly population in Taiwan with respect to the available resources in the long-term healthcare sector.

範例 iii.

Therefore, we recommend devising a credit risk assessment model by analyzing a mass volume of data or detecting concealed purchasing models, thus reducing the defaulted loan burden of small financial institutions.

範例 iv.

Therefore, we recommend developing an AHP-based method that enables administrators of senior citizen residential communities to identify critical factors for successful operations.

範例 v.

Therefore, we recommend developing a patient selection model using AHP and TOPSIS, which analyzes incoming hospital patients with respect to the public image of the hospital, the current market demand for specific services, as well as efforts to gain the confidence of patients in the quality of medical services.

範例 vi.

Therefore, we recommend developing an inventory stock management model that coordinates the efforts of suppliers and retailers, in which the inventory stock model of distribution for the normality is adopted to determine the minimum total cost via the metering method.

C.方法論：您的機構計畫如何實行這個目標？

範例 i.

Based on numerous customer data available, a data mining method, i.e., CRISP-DM, can be adopted, which combines the conventional means of data exploration with two mathematical calculations (decision tree and category nerve) to determine how various purchasing activities are related and how many factors can rank the value of a customer's relationship. Among these factors include the types of products purchased, their quantity, their net interest of margin and their cost of acquisition and servicing. Factors associated with customer relations and customer life cycle can then be integrated to construct an enhanced management model.

範例 ii.

A multi-regression model can be developed to measure and forecast not only the quantity of early demand, but also the relationship between the demand and critical factors by accumulating a significant amount of data and identifying such factors. The data can then be collected via a questionnaire, with the factors quantitatively measured based on a method found in literature review. Next, a GM (1, 1) model based on the Grey Theory can be developed to accurately forecast the supply of available long term healthcare resources by using data acquired from the website of Taiwan's Ministry of Interior.

範例 iii.

A database can be constructed, capable of integrating customer data, in which highly effective data mining approaches can be applied to identify the attributes of each customer account, such as overdraft records, outstanding loans and income level. A credit ranking criteria based on a decision tree solution can then be established for all customers in the bank database.

範例 iv.

A questionnaire survey can be designed to select the most appropriate sampling method. A valuation standard that reflects enterprise values can then be derived based on interviews with field experts. Next, critical factors for successful operations can be identified using the AHP method.

範例 v.

AHP and TOPSIS can be adopted to analyze incoming hospital patients with respect to the public image of a hospital, current market demand for specific services, efforts to gain the confidence of patients in the quality of medical services and factors why patients select a certain hospital for treatment. A frame questionnaire can then be designed by incorporating the basic requirements of patients, including intangible ones such as respect. Next, the following criterion can be adopted to analyze the questionnaire results: physicians are qualified and reliable; medical staff has received sufficient training; physicians are willing to offer valuable clinical information; physicians express concern for their patients; medical staff display a professional attitude; and the hospital honors patient confidentiality. Additionally, SPSS software package can be used to analyze those results. Moreover, four assumptions can be raised: a)Public appraisal over the quality of medical treatment affects the overall results; b)The degree of service quality influences the expectations of hospital patients; c)Patient's demand influences the service quality that hospital patients expect; and d)A patient's experience influences the service quality expected.

範例 vi.

The (s, Q) policy of the inventory stock between the supplier and the manufacturer can be adopted to reduce the amount of minimum stock required to reduce the total cost when ordering large quantities. Problems encountered between the supplier and the manufacturer in predicting the required amount of inventory stock can be addressed individually. Next, the stock model (s , Q) can be utilized to derive the inventory stock models for the supplier and manufacturer without complex mathematical equations For the inventory stock cost, sales volume and orders in short supply, the most appropriate control channel can be adopted by the supplier and the manufacturer, followed by mathematical analysis and estimation.

D.預期的結果：達成目標的立即利益為何？

範例 i.

Based on the proposed model, factors of the ranking module can be verified and adjusted to ensure that a company continuously provides quality services. While customer's value in customer relationship management can be determined, this model can significantly enhance the ability to attract new customers.

範例 ii.

In addition to identifying relevant factors that affect the quantity of demand, the proposed forecasting models can also measure precisely the quantity of demand and supply of long-term healthcare resources in Taiwan.

範例 iii.

In addition to greatly facilitating the decision of a banking officer on whether to grant a loan, the credit risk assessment model can reduce operating costs by enhancing the process flow.

範例 iv.

The proposed AHP-based method can enable enterprises involved in the development of senior citizen residential communities to identify the features of and prerequisites for successful operations.

範例 v.

The proposed patient selection model can be established using AHP and TOPSIS, which conventional statistical models fail to do. While influencing which hospital that a patient selects, quality of hospital services does not necessarily improve customer satisfaction.

範例 vi.

The proposed inventory management model can determine the most economic purchasing amount, purchasing opportunity and inventory stock policy for both the supplier and manufacturer.

E.全面性的貢獻：達成目標後會對管理界其他人造成何種利益？

範例 i.

Furthermore, this model can be used in other business sectors to enhance the ability to identify, acquire and retain loyal and profitable customers.

範例 ii.

The proposed forecasting models can provide a valuable reference for health care policy makers, investors in the medical sector, administrators and academics when devising relevant policies and strategies.

範例 iii.

Moreover, the proposed model can pave the way for other potential data mining applications in financial institutions, such as more thoroughly satisfying customers through more effective marketing strategies based on acquired data. The proposed model is highly promising for other industrial applications as well.

範例 iv.

The proposed method can enable administrators of senior citizen residential communities to select the most feasible options during decision making by considering finance and market analysis related concerns in order to ensure success in daily operations. In addition to enhancing the competency of administrators in making their residential communities productive in this increasingly competitive sector, the proposed method can provide a valuable reference for experts, academics and investors in the long term sector when attempting to devise relevant marketing strategies.

範例 v.

Moreover, the proposed model provides an effective strategy for hospitals attempting to generate revenues independent of the national health insurance scheme.

範例 vi.

By using random and the most appropriate control measures, the proposed management model can determine precisely the demand between the supplier and the manufacturers, reduce overhead costs and enable enterprise managers to select the most effective inventory stock policy to enhance overall operations.

【參考文獻】

柯泰德（2002）。《有效撰寫英文工作提案》。台北：揚智。

Unit Two

Writing Effective Workplace Memos: Persuasive Reports

有效撰寫管理英文備忘錄——說服力的展現

簡 介

工作備忘錄是台灣管理師和國外相對合作對象溝通的必需文件，本單元描述這種常見的工作備忘錄。為了使管理類的讀者採行計畫或策略，您須遵循以下步驟：首先引起讀者注意，再者建立讀者對計畫、策略或服務的需求，展示計畫、策略或服務的可行性，引導讀者預想計畫、策略或服務實行後的成果，最後採取特定行動。台灣管理案例在此也一併提供。

A.引起讀者注意：描述一個事實或情勢跟管理類的讀者所關心的特定情勢做連結

範例 i.

Globalization of the cosmetics sector is widespread in most industrialized countries. Taiwan's biotechnology industry has rapidly evolved in recent years, subsequently increasing marketing efforts in the local cosmetics sector, which originally focused only on healthcare products. For instance, according to the Industrial Technology Research Institute, revenues from the local cosmetics sector ranged between $US 1,700,000,000 and $US 1,800,000,000 in 2002. This figure contrasts with $US 700,000000 for the health food market and $US 900,000000 for the food testing experiment market. Many female consumers regard cosmetics as a daily necessity, as reflected by the wide array of cosmetic brands in the competitive market.

範例 ii.

Complex administrative procedures within Taiwan's National Health Insurance (NHI) scheme have led to errors in insurance claims and much inefficiency. For instance, adhering to all NHI regulations would require filling out more than thirty forms, depending on various commercial demands, e.g., insurance claims, salary readjustments and modifications of insurer's/claimant's name.

範例 iii.

Although the Taiwanese government has implemented separate frameworks for the medical and pharmaceutical sectors since 1997, the amount of hospital prescription drugs filled by pharmacies is extremely low. This trend explains why over-the-counter drug purchases, along with sanitary and other related medical products, have become major revenues for pharmacies. Local pharmacies have thus adopted a business model of one-stop shopping in recent years in which customers can purchase a wide array of medical and health food products.

範例 iv.

The extremely competitive medical market sector in Taiwan and budget deficits incurred from the island's National Health Insurance scheme have led to the implementation of a Global Budget System. Whereas hospitals heavily prioritize attracting new patients, the growing patient turnover rate has received increasing attention. Thus, hospitals must concentrate their operational efforts on retaining current patients and reducing the turnover rate of those going to other medical centers for treatment.

範例 v.

While facing an increasing elderly population, Taiwanese must effectively address the increasing demand for long-term healthcare facilities and services. The potential growth for this non-profit market sector, the largest one in Taiwan, is immense, making it impossible for healthcare providers to resist expansion into this area. Therefore, forecasting the long-term care market trends (both in supply and demand) is a fundamental aspect of feasibility analysis. Both governmental policymakers of social welfare trends and commercial investors heavily rely on forecasting reports to remain abreast of regulations governing health finance policies and to develop inventory projects.

範例 vi.

Lending and investing are common business practices of financial organizations. Given that most Taiwanese enterprises are small- and medium-sized, their credit rating status is unknown. Banks have thus expended considerable personnel and time in diagnosing the ability of enterprises to repay loans in order to reduce the likelihood of non-performing loans (NPLs). Further exacerbation of the NPL crisis would incur a shortage of operating funds in banks, ultimately harming investor interests, stockholder profits and institutional reputation.

B.建立讀者對計畫、策略或服務的需求：描述管理問題和需求間的連帶關係，以及不解決問題的後果

範例 i.

Given the enormous female consumer demand and intense competition, manufacturers must more thoroughly understand the consumer purchasing behavior of this market niche. Cosmetic manufacturers are concerned not only with marketing practices, but also with the potential lowering of product and service quality, necessitating the adoption of the 4P method. Although receiving considerable attention in conventional marketing, the 4P method has seldom been explored with respect to its effectiveness in cosmetics marketing practices in order to enhance the competitiveness and market share of related manufacturers. The failure of cosmetics manufacturers to adopt the 4P method in their marketing practices will make them less competitive in expanding their market share as well as promoting current products and services.

範例 ii.

With errors that confused applicants make when filling out appropriate forms, the NHI staff spends much time in handling errors and then requesting insurers and claimants to amend erroneous information. Despite the enormous amount of administrative time and costs involved in handling these errors, this growing concern and its larger implications have not been addressed in previous literature. The inability to gradually reduce the amount of human resources involved in handling insurer and claimant errors and simplify administrative procedures as well as NHI forms will lead to higher operational costs owing to errors such as those made on the identification card, e.g., mistakes in cardholder number, birthday, name or filling out of the wrong form. In practice, telephone, fax or mail is used to correct such errors, subsequently creating substantial overhead costs, as well as additional time spent in human resources. Given the limited number of available NHI staff, simplifying forms and procedures is of priority concern.

範例 iii.

However, the inability to accurately forecast medical and health care share of household consumption and the number of local pharmacies to be established makes it nearly impossible for managers to analyze market competition precisely and develop effective strategies.

範例 iv.

Incapable of determining patient turnover rates precisely, hospitals will expend considerable costs in attracting new patients and reduce the subsidized revenues from the National Health Insurance scheme, ultimately lowering their competitiveness.

範例 v.

Whereas the modeling method has seldom been adopted to forecast market trends in long-term care in Taiwan, most studies focus mainly on the expert conjecture method or the principle of relative proportion of previous growth to forecast market supply and demand. Those methods not only neglect the market and the factors related to the long-term healthcare sector, but also result in individual and organizational estimates in which a high degree of variance exists for the demand.

範例 vi.

Banks loan to or invest in organizations based on whether they have a good credit rating, which involves investigations undertaken by financial experts examining financial reports to determine current status and classify credit results. Such time consuming and prohibitively expensive investigations do not always yield satisfactory results and even miss market opportunities. The inability of loaning institutions to accurately estimate the credit rating of enterprises will incur misclassifications that lead to inestimable losses and prevent machine learning methods to yield a higher accuracy in expert systems owing to the inability to accurately predict credit ratings.

C.展示計畫、策略或服務的可行性：詳細描述完成計畫或策略目標的必須步驟

範例 i.

To resolve the above dilemma, we have developed a novel 4P-based marketing strategy for the local cosmetics sector within Taiwan's biotech industry by considering consumer needs under the principles of product, price, promotion and place. This strategy can enable managers in the cosmetics sector to more thoroughly understand consumer preferences not only by learning how to identify and target potential customers efficiently, but also by establishing a retention strategy to maintain loyal customers and attract new ones. A questionnaire is submitted to cosmetic manufacturers on the most appropriate marketing method, followed by factor analysis of those results associated with such an approach that adopts 4P principles.

範例 ii.

To address this need, we have developed a novel administrative procedure for classifying and simplifying forms to pay insurance premiums and file other insurance-related claims in Taiwan's National Health Insurance (NHI) scheme. A database containing detailed information of NHI insurance holders is accessed while ensuring the confidentiality of such information, thus reducing administrative costs and the number of personnel involved. Three insurance claim-related forms are developed to reduce the redundancy of the more than thirty existing ones, thereby reducing data processing significantly. A highly restricted networked-based system is then accessed to ensure the confidentiality of NHI customer data and also reduce the amount of paperwork that insurance holders must submit. Accessing the system would require staff setting up an account and logging into the system during use, with a systems administrator assigned to maintain the smooth flow of operations. The latter is especially important given the extremely heavy daily workload in which staff must handle significant amounts of data daily and have no time to maintain system operations.

範例 iii.

To compensate for this limitation, we have devised a feasible forecasting method to estimate the growth of medical and health care expenditures as well as pharmaceutical units in Taiwan. Medical and health care expenditure-related data from 1999 to 2002 are obtained from the Central Region Office of the Budget, Accounting and Statistics, Executive Yuan. Data on pharmaceutical units are then obtained from the 2003 Annual Report of the Department of Health. Based on data from those sources, the GM (1, N) model of the Grey theory is applied for forecasting purposes.

範例 iv.

Based on the above, we have developed a novel prediction model to identify the turnover rate of customers in the medical sector. More specifically, the proposed model can incorporate a novel set of management strategies that emphasize customer retention within the medical sector to strengthen a hospital's competitiveness. A database containing pertinent hospital patient data is utilized by applying a data mining method to identify the factors associated with customer turnover rate. A data mining approach that incorporates various data analysis tools is also adopted to discover interesting trends and relationships among various data sets. Pertinent literature is then reviewed to confirm the reliability of variables in the database. Next, questionnaires are sent to hospital administrators regarding the level of customer satisfaction. Additionally, distinct consumer groups are identified by performing cluster analysis to distinguish between all consumer groups. Moreover, neural networks are used to enhance the model accuracy, with those results subsequently analyzed.

範例 v.

Therefore, we have developed two forecasting models to estimate the market demand of the elderly population in Taiwan with respect to available resources in the long-term healthcare sector. A multi-regression model is developed to forecast not only the quantity of the elderly population in Taiwan, but also the relationship between the elderly population variation and critical factors. Following the submission of a questionnaire to institutional managers as well as public and private sector investors, critical factors associated with commitment to further develop Taiwan's long term healthcare sector are identified from those questionnaire replies based on pertinent research from a literature review. A GM (1, 1) model based on the Grey Theory is then developed to accurately forecast the supply of Taiwan's long term care facilities and services by using data acquired from the Ministry of Interior's website.

範例 vi.

To resolve this problem, we have developed a novel classification model for small- and medium-sized enterprises, capable of increasing the prediction accuracy of classifying the credit ratting of enterprises and reducing administrative expenses. Data from small- and medium-sized enterprises are accumulated and categorized, followed by processing with a fuzzy set to store additional information. Fuzzy data are then clustered using a two-stage clustering approach to obtain a proper classification. Next, an ANN structure is constructed by using the cluster results as an output of ANN. Additionally, the ANN classification machine is estimated with 10-fold cross validation to identify the most efficient machine. Moreover, results obtained from the above tests are tabulated and compared with those in literature to verify data authenticity.

D.引導讀者預想計畫、策略或服務實行後的成果：強調採行該計畫或策略後立即和長期的利益

範例 i.

Adopting this 4P-based marketing strategy can enable local cosmetic manufacturers to increase their market share in related products and services. The proposed strategy also provides a more objective outcome for members when making related decisions than conventional approaches can, thus accelerating the marketing process. In addition to encouraging product innovation, the 4P-based marketing strategy can provide Taiwan's biotech industry with clear guidelines on how to equip management in the local cosmetics sector with appropriate and efficient marketing policies that will ultimately decrease operating costs and enhance competitiveness. Moreover, the proposed method can reveal how the biotechnology industry incorporates 4P concepts, thus clarifying the behavioral patterns of cosmetics customers.

範例 ii.

The proposed administrative procedure streamlines the process of filing insurance-related claims by simplifying the form procedure and enable access to relevant customer data on line in an efficient manner. Importantly, the proposed procedure significantly reduces not only the amount of time to fill out insurance claims, but also postage fees owing to the ability of the network-based system to accelerate data processing. This highly restricted networked-based system that would ensure NHI customer confidentiality is unique in that it not only integrates the efforts of various organizations within the NHI scheme by providing data access, but also makes NHI administrative services more flexible for customers.

範例 iii.

The proposed forecasting method can accurately estimate medical and health care expenditures as well as the number of pharmacies to be established in Taiwan from 2008 to 2010. The proposed method can also identify how the medical and health care share of household consumption is related to the number of pharmacies. Importantly, the proposed method provides a valuable reference for both governmental authorities in formulating policies and pharmaceutical managers in developing competitive marketing strategies.

範例 iv.

In addition to providing the medical sector with more accurate guidelines on patient retention and marketing planning, the proposed model can greatly facilitate hospitals in providing high quality and flexible health care services that will ultimately enhance their public image with markedly improved relations with patients. More than identifying the major factors underlying customer turnover rate, the proposed model can also offer feasible strategies to cope with this dilemma and achieve management goals. Furthermore, the proposed model can contribute to efforts to maintain customers in the highly competitive medical market sector as well as provide a valuable reference for healthcare managers in enhancing customer relations.

範例 v.

In addition to identifying relevant factors that affect the quantity of the elderly population, the proposed forecasting models can also measure precisely the quantity of demand and supply of long-term healthcare resources in Taiwan. Moreover, the proposed models can provide not only a valuable reference for institutional managers, private sector investors and academics, but also a feasible strategy for health care policy makers when devising strategies for fulfilling the demand of this rapidly growing sector.

範例 vi.

The proposed classification model can increase prediction accuracy from 10% (as achieved by the conventional classification system) to 80%, significantly reducing operational costs. In addition to increasing the efficiency of obtaining data from small- and medium-sized enterprises and reducing operational costs during estimation processing, the proposed model provides promising guidelines for banks.

E.最後採取特定行動：管理類的讀者經由完成計畫或策略來有效貢獻管理部門中的特定領域

範例 i.

We seek your expedient approval to implement this novel marketing strategy in your organization so that you can more easily understand behavioral patterns of your cosmetics customers

範例 ii.

We therefore highly recommend that you adopt this administrative procedure to streamline the processing of insurance premiums and other insurance-related claims in Taiwan's National Health Insurance (NHI) scheme.

範例 iii.

We therefore seek your authorization to adopt this forecasting method to ensure that an adequate number of pharmacies located strategically can comply with health consumer demand in Taiwan.

範例 iv.

We seek your approval to adopt this customer retention strategy in order to reduce the operational costs of hospitals and increase the number of overall patients.

範例 v.

We strongly recommend that your healthcare organization adopt these forecasting methods not only to determine the accuracy of forecasting reports, but also to determine what policies or inventory projects to implement.

範例 vi.

We therefore highly recommend that you adopt this novel classification model to increase your accuracy of classifying the credit ratting of lending enterprises and significantly lower administrative expenses.

【參考文獻】

柯泰德（2002）。《有效撰寫英文工作提案》。台北：揚智。

Unit Three

Writing Effective Workplace Memos: Informal Technical Reports

有效撰寫管理英文備忘錄——非正式實用管理技術報告

簡 介

工作備忘錄是台灣管理師和國外相對合作對象溝通的必需文件，本單元描述常見的工作備忘錄：非正式實用管理技術報告。有效的管理技術報告提供精確的訊息以供決策的制定。內容包括工作備忘錄裡的資料解讀、管理方案評估、結論及建議的形成。管理技術報告同時告知影響機構的情勢，例如顧客的偏好及目前可得的市場。成功的管理技術報告隨同台灣管理方案案例也一併提供。

A. 簡要的描述管理方案所關心的事項，經由一個句子描述管理機構對影響目標工業或客戶有關事項的關心程度

範例 i.

Our project addressed how to identify major success factors of the pharmaceutical sector in Taiwan for devising more effective customer-based strategies.

範例 ii.

Our recent effort addressed how to determine accurately the customer turnover rate in our hospital.

範例 iii.

Our group has thoroughly explored the difficulty of forecasting the supply of and demand for long-term healthcare in Taiwan.

範例 iv.

The bank' s board of directors expressed concern over how to identify, acquire and retain loyal and profitable customers by more effectively managing customer relations.

範例 v.

Our recent effort addressed how to adopt marketing practices in our hospital' s public relations department to enhance our competitiveness in the medical sector.

範例 vi.

Our working group has become increasingly intrigued with how to devise appropriate strategies and procedures for implementing customer services based on available client data.

範例 vii.

Our recent project addressed how to select the optimal location and size of Taiwanese correctional facilities, thus ensuring a design that minimizes societal risks.

B.闡明特定部門或客戶所關心的工業環境：管理問題形成的工業環境背景描述

範例 i.

The Taiwanese government has relaxed restrictions on the domestic medical sector in recent years, such as enabling consumers to purchase over the counter drugs, vitamins and other medicines in supermarkets or through mass merchandisers. Consequently, the profits of local pharmacies have been negatively impacted.

範例 ii.

The extremely competitive medical market sector in Taiwan and budget deficits incurred from the island's National Health Insurance scheme have led to the implementation of a Global Budget System. Whereas hospitals heavily prioritize attracting new patients, the patient turnover rate has received increasing attention.

範例 iii.

Although Taiwan's growing elderly population has increased the demand for institutional-based health care, the institutional-based health care sector is obviously disproportional island wide in the level and quality of services and facilities, resulting in significant public concern over this dilemma.

範例 iv.

Both the increasing popularity of credit card use and growing number of Internet-based promotional activities in Taiwan have enabled banking institutions to acquire extensive customer data.

範例 v.

Governmental policy over Taiwan's National Health Insurance (NHI) scheme continuously changes, especially in light of increasing medical costs and premiums as well as concern over the potential lowering of health care quality offered. Hospitals must thus emphasize marketing practices to attract patients through public relations.

範例 vi.

As a governmental subsidized, non-profit research organization, Industrial Technology Research Institute (ITRI) has established a hot line to handle customer concerns in order to more fully comply with the customer oriented concept. Via a unique phone line in the call center at ITRI, customers can query about or request various services, as well as express their dissatisfaction with service quality. In response, ITRI provides an applicable procedure to process these calls. The hot line currently receives an average of 150-200 monthly, with a total of 4016 calls made domestically up until August 23, 2004.

範例 vii.

Given the stagnant Taiwanese economy and increasing unemployment levels in recent years, growing crime rates and subsequent convictions have led to a high prisoner population in the island's correctional facilities.

C.介紹管理問題：管理問題的本質和對與其相關的特定部門或客戶的負面影響

範例 i.

Although pharmacies have attempted to strengthen consumer demand through one-stop shopping, e.g., selling sanitary and other related medical products, overall revenues have not significantly increased.

範例 ii.

Of the relatively few studies that have addressed the medical sector in this area, conventional approaches calculate the number of patients on a daily basis without differentiating between new and return patients.

範例 iii.

Whereas concern over the quality of institutional-based care and the working capital weight correlation has seldom been addressed, most studies focus mainly on legislation, human resource management and approaches to enhance patient care.

範例 iv.

In addition to helping banking institutions to execute customer management and service management efficiently, thoroughly analyzing such data can optimize marketing management practices. Corporate survival in the future hinges on the ability to know and treat customers well by analyzing pertinent data.

範例 v.

Although the role of public relations in business marketing has received considerable attention, its role in hospital marketing practices has seldom been addressed.

範例 vi.

The call center generally focuses on integrating IT and communication technologies through a database that records all information on customer interactions. However, customer value is seldom analyzed nor the performance of the service strategies evaluated.

範例 vii.

However, the conventional means of selecting the locations of correctional facilities lacks an objective approach, often depending on the subjective judgment of decision makers.

D.介紹管理方案的目標：管理機構對以上問題最合理的回應

範例 i.

Therefore, we developed a key successful factors (KSF) model that incorporates consumer purchasing factors on different hierarchical levels so that local pharmacies in Taiwan can reform their marketing strategies.

範例 ii.

Therefore, we developed a novel prediction model to identify the turnover rate of customers in the medical sector and determine how to retain them.

範例 iii.

Therefore, we analyzed how various working costs impact the quality of institutional-based healthcare.

範例 iv.

Therefore, our research group developed a customer ranking model capable of analyzing the dynamic purchasing behavior of customers and identifying the potential ones who can generate bank revenues.

範例 v.

Therefore, we launched a marketing strategy for our hospital that emphasizes identifying and satisfying patient needs from a customer' s perspective.

範例 vi.

Therefore, we examined previously unrecognized customer behavioral patterns. Based on those results, we further recommended appropriate strategies and procedures for implementing customer services.

範例 vii.

Therefore, we attempted to identify those factors impacting the infrastructure and safety of correctional facilities in Taiwan via the analytic hierarchy process (AHP) to determine the optimal target population and location for such facilities.

E.管理方案方法論的細節描述：精確的方案步驟描述

範例 i.

A questionnaire based on consumer purchasing factors identified from pertinent literature was sent to consumers and pharmaceutical managers. Major consumer purchasing factors were then analyzed and ranked using the analytic hierarch process.

範例 ii.

A database containing pertinent hospital patient data was utilized by using a data mining method to identify the factors associated with customer turnover rate. Pertinent literature was then reviewed to confirm the reliability of variables in the database. Next, questionnaires were sent to hospital administrators regarding the level of customer satisfaction, with those results subsequently analyzed.

範例 iii.

Exactly how institutional-based healthcare quality and working costs correlate with each other was analyzed by using data of financial statements and customer satisfaction. Relevant data (not in a digitized form or in small quantity) was then calculated by using a gray system-based mathematic method and fuzzy theory

範例 iv.

Those behavioral results were used to devise diverse promotional strategies or customize products or services according to consumer needs, thus achieving market differentiation and effective management of customer relations. Via the proposed model, factors of the ranking module were verified and adjusted to ensure that a company continuously provides quality services.

範例 v.

Factor analysis of customer needs was performed through personal interviews. The results were then analyzed, along with vital factors identified as well.

範例 vi.

Data was initially retrieved from the customer service system and then categorized sequentially. Areas to be analyzed were then assigned to research team members. Next, customer attributes and the category to which customer queries belong were analyzed according to the assigned topics. Finally, appropriate strategies and procedures for implementing customer services were recommended

範例 vii.

Location-related factors were identified through an exhaustive literature review and consultation with experts in the field. Exactly how these individual factors are related to each other was also identified through use of AHP. Next, a questionnaire was submitted to administrators of correctional facilities to examine the correct factors. Additionally, all factors involved in selecting the target population and location of correctional facilities were analyzed using AHP, with the optimal location chosen based on those factors.

F.管理方案主要成果總結：管理方案對特定部門或客戶立即的利益

範例 i.

The proposed KSF model can enable pharmaceutical managers to execute business operations more effectively by allowing them to modify marketing strategies based on how consumer demand and the response to those demands differ.

範例 ii.

The proposed prediction model can be adopted to design and implement precautionary measures towards customer turnover rates in related fields.

範例 iii.

After the fuzzy theory is applied to digitize the data, the gray system can be used to rank and verify the importance of various working costs and the quality of institutional-based healthcare.

範例 iv.

The proposed customer ranking model can be adopted to effectively manage customer relations, thereby reducing promotional costs significantly and allowing sales staff to concentrate on identifying potential customers. While the customer's value in customer relationship management can be determined, this model significantly enhances the ability to attract new customers.

範例 v.

The proposed marketing strategy can increase customer satisfaction by over 10% through hospital marketing by significantly reducing overhead costs and work time.

範例 vi.

While combining qualitative and quantitative approaches, the content analysis approach can not only transfer the qualitative data into quantitative data for statistical analysis, but also explore the qualitative implications via an encoding process.

範例 vii.

The optimal target population and location for future correctional facilities are identified based on analysis results. Meanwhile, a newly established correctional facility can alleviate concerns of nearby residents by reassuring them of the security features and pointing out potential economic benefits. While correctional facilities operate under a governmental budget, selecting an optimal location can economize facility operations and maintenance-related management costs.

G.管理方案對特定部門或領域的全面貢獻：研究結果和所提方法對管理機構以外更廣大讀者的牽涉

範例 i.

Importantly, the proposed model can also be applied to other retail stores, subsequently enhancing their business operations.

範例 ii.

In addition to identifying the major factors underlying customer turnover rate, the proposed model can offer feasible strategies to cope with this dilemma and achieve management goals. Moreover, the proposed model contributes to efforts to maintain customers in the highly competitive medical market sector and provides a valuable reference for healthcare managers in enhancing customer relations.

範例 iii.

Results of this study can enable institutional-based healthcare facility managers not only to strengthen areas of corporate management, but also to provide a high quality of long term healthcare.

範例 iv.

Furthermore, this model can be used in other business sectors to enhance the ability to identify, acquire and retain loyal and profitable customers.

範例 v.

The proposed marketing strategy provides a valuable reference for hospital managers attempting to develop an optimization procedure for implementing marketing practices precisely.

範例 vi.

Furthermore, this method can effectively analyze non-structural data.

範例 vii.

Results of this study have provided a valuable reference for governmental authorities in selecting the optimal location and size of correctional facilities, thus ensuring a design that minimizes societal risks. Furthermore, this optimal location strategy can encourage other law enforcement organizations to understand how location planning can ensure the security of an inmate population, reduce societal risks and alleviate public concern by making local residents aware of the potential benefits of such facilities.

【參考文獻】

柯泰德（2002）。《有效撰寫英文工作提案》。台北：揚智。

Unit Four

Effectively Communicating Online

有效海外管理英文交流——探討管理師如何有效的與海外專業人士交流英文

簡 介

本單元主要教導國內管理師如何建構以下相關專業英文信函，以用來和國外管理相關專業人士有效溝通。內容包括：A.科技訓練請求信函：某人學歷及工作經驗概述、科技訓練申請、讚揚提供科技訓練的機構、科技訓練細節解釋及表達對科技訓練機會的樂觀態度。B.資訊交流信函：資訊流通關係的提呈、為資訊流通而提供個人或機構經驗、為資訊流通而讚揚某機構或某人、資訊流通關係的細節詳述、表達對資訊流通關係的樂觀態度及開始資訊流通關係。C.科技訪問信函：科技訪問許可申請、闡述科技訪問的目地、介紹某人的能力或機構經驗、對訪問機構的讚揚、詳述科技訪問的細節及科技訪問核准的確認申請。D.演講者邀請信函：邀請演講者、描述演講目地、詳述演講的細節、詳述交通及住宿的細節、演講者行前必知指導及對演講者強調訪問的重要性。E.旅行安排信函：行程細節列表及對行程細節列表的建議。F.資訊請求信函：解釋請求資訊的目地及說明學術專業經驗。

A. Seeking Technical Training / 科技訓練請求信函

1.Summarizing one's academic or professional experiences / 某人學歷及工作經驗概述，請參考以下範例：

The attached resume and recommendation letters provide further details of my solid academic background and professional experiences.

Relevant project experiences have significantly strengthened my independent research capabilities as well as statistical and analytical skills.

Intensive on-the-job training enabled me to handle various requests and difficult situations, enabling me to resolve problems efficiently.

Performing independent research has definitely made me more receptive to accepting new challenges.

In addition to these academic and professional activities, I often read supplemental literature to remain abreast of the latest technological trends.

第一個練習：

以下每句爲用來訓練有關科技訓練請求信函：某人學歷及工作經驗概述，請把左邊及右邊的每一片語連結成一個完整句子。解答在本章最後。

1. My graduate school research focused on designing and applying relevant marketing strategies to the industrial engineering sector,	a. I will complete my master's degree requirements from the Institute of International Finance at National Cheng Kung University in the spring of 2010.
2. First exposed to business management during undergraduate school,	b. the need to continuously grow academically and professionally, albeit with a cordial and earnest attitude.

3. I received my undergraduate training in the Department of Business Management at National Cheng Chi University,	c. personal resume, university academic transcripts and letter of recommendation from our university president.
4. The attached resume will give you a better idea as	d. as evidenced by my ranking among the top five scholastically in related courses.
5. University instructors equipped me with strong academic fundamentals in business management,	e. to how my personal strengths match the needs of your research group.
6. I view myself as a responsible individual that highly prioritizes	f. which further spurred my interest in this area of research.
7. Please find enclosed my	g. with a subsequent Master's degree in Business Management from Dong Hwa University.
8. Marketing has interested me immensely throughout undergraduate and graduate school,	h. with those findings submitted to an international journal for publication.

2.Requesting technical training／科技訓練申請，請參考以下範例：

My academic advisor, Dr. Cheng, recommended that I contact you regarding the possibility of a guest researcher stay in your laboratory.

Eager to strengthen my knowledge expertise in this field, I hope to serve in your laboratory as a guest worker to compensate for my lack of training and practical laboratory experiences in this area.

Eager to understand the dynamic environment that your work involves, I would like to arrange for a three-month stay in your laboratory as a self-supported guest researcher, hopefully this upcoming summer.

The opportunity to serve in a self-supported guest researcher position in your laboratory would further orient me on how to become more proficient in this line of research.

To build upon the above academic and professional experiences, I would like to serve as a self-supported guest researcher in your company's Strategic Planning Department for a three month period, hopefully during my upcoming summer vacation.

3.Commending the organization to receive technical training from / 讚揚提供科技訓練的機構，請參考以下範例：

As your laboratory has devoted much time in researching the above topics, I would greatly benefit from the opportunity to work in your laboratory as a self-supported researcher during my upcoming summer vacation.

As a leader in this field of research, your laboratory would enable me to further refine my professional skills in strategic management.

While searching for pertinent research literature in my field, I often come across your findings published in several reputed international journals.

As is well known, your research group has devoted considerable time and resources to researching business management-related topics.

From your laboratory's online introduction, I am impressed with the comprehensive strategies that you adopt in the workplace.

第二個練習：

以下每句爲用來訓練有關科技訓練請求信函：讚揚提供科技訓練的機構。請把左邊及右邊的每一片語連結成一個完整句子。解答在本章最後。

1. After thoroughly reviewing your online literature and promotional materials, I am especially drawn to the marketing strategies that your organization has adopted,	a. in holding the 12th International Conference on Business Management, I am most interested in learning from your rich experience in this area.
2. Having read upon your organization's research findings in the proceedings of several international medical conferences and globally renowned journals, I also am anxious	b. strategic management research for quite some time, developments that I have closely followed for quite some time.
3. Given your organization's recent success	c. as evidenced by the publishing of your research findings in several internationally renowned journals.
4. I am especially drawn to your corporation's emphasis on customization that	d. the most appropriate place.
5. Your country has pioneered	e. as well as its globally recognized products and services for improving the lives of urban dwellers.
6. Your company has distinguished itself in	f. in researching finance-related topics.
7. As is well known, your research group has made considerable progress in adopting the latest international finance approaches,	g. orients consumers on the reliability of your products and services.
8. Your division is widely recognized as a leader	h. progressive policies in welfare of its elderly population, including the construction of state-of-the-art facilities.

9. Given my research interests, your laboratory appears to be	i. explaining why I am eager to work in your organization as a guest researcher during my upcoming summer vacation.
10. I hope to perform advanced research in your center owing to its highly skilled personnel, state-of-the-art equipment and facilities,	j. which are quite compatible with my current esearch interests.

4.Explaining the details of the technical training / 科技訓練細節解釋，請參考以下範例：

Specifically, we are concerned with how to effectively adopt the latest business management approaches.

If given the opportunity, I would like exposure to the following topics:

As for my accommodation during the guest stay, I would like to lease a single-room hotel, at a range between US$150 to US$200 daily.

If granted this opportunity, I am open to investigating related topics with your highly skilled staff.

If possible, I hope to work directly under your supervision.

5.Expressing optimism in the possibility of technical training / 表達對科技訓練機會的樂觀態度，請參考以下範例：

I look forward to hearing from you soon as to whether such cooperation would be feasible.

I am confident in my ability to work under your supervision in a research fellowship with your

highly skilled research personnel, an opportunity that would greatly advance my competency as a financial planner.

The opportunity to work in a practicum internship in your company would provide me with an excellent environment not only to fully realize my career aspirations, but also to upgrade my own technological expertise.

Exposure to the way in which your department applies the latest business management concepts in a practical context would give me a clearer direction not only my graduate level research, but also my chosen career path as well.

My previous experience in applying theoretical concepts to analysis and related applications will hopefully prove to be an asset to any effort that I belong to in your laboratory.

第三個練習：

以下每句爲用來訓練有關科技訓練請求信函：表達對科技訓練機會的樂觀態度，請把左邊及右邊的每一片語連結成一個完整句子。解答在本章最後。

1. Somewhat familiar with the scope of your company's financial activities, I am confident that my previous	a. matches that of your research group, I am anxious to look for potentially collaborative activities that would be mutually beneficial.
2. Such an opportunity would compensate	b. to any collaborative effort that I belong to in your company.
3. I would greatly appreciate your comments	c. regarding this proposal.

4. I am confident that my previous academic and professional experiences will prove valuable	d. context for the concepts learned in class.
5. I bring to your corporation a solid understanding of	e. academic and professional experiences will prove valuable to any collaborative effort that I belong to.
6. Given that the research direction of our laboratory closely	f. as an aspiring researcher in this field, as I search for a more practical context for the academic concepts taught in class.
7. Working in your organization would give me a practical	g. research design and methodology that will hopefully contribute to your company's efforts to enhance operational quality.
8. The opportunity to work in your laboratory would mature me	h. in applying theoretical concepts to survey research and statistics.
9. More than the opportunity to receive technical training, this four month technical visit	i. for my lack of formal professional experience in this area.
10. I hope that I can contribute my previous experience	j. will hopefully initiate a long-term collaborative relationship between our two laboratories.

B. Sample Professional Training Application Letters / 專業訓練申請信函

Dear Miss Preston,

My academic advisor, Professor Hsu, suggested that I contact you on the possibility of strengthening my professional skills in human resource management by working in a self-supported capacity in your company for six months. As a graduate student in the Master's degree program in Business Administration at National Taipei University, I also have come into contact with information management, technology management and engineering management in my previous studies. Still, I lack the necessary skills in human resource management, an area in which your company has acquired much expertise. The opportunity to work in your company would allow me to apply the academic concepts in class to a practical setting.

Under the guidance of my academic advisor Professor Hsu Pi Fang, I am especially interested in the human resource managerial aspect of Yahoo! Taiwan. Well aware of your proficiency as a human resource department manager, I am impressed with your distinctive administrative approach, as evidenced by your numerous accomplishments. Given my undergraduate background in science, I wish to compensate for my lack of human resource management skills through this work opportunity under your direct supervision. I hope to gain as much exposure as possible to personnel-related problems that arise daily in a large enterprise such as yours. I am especially interested in how your company cultivates employee potential through company policies and training programs. The opportunity to receive training through practical exposure to your unique organizational culture will definitely benefit my graduate research.

Please carefully consider my request. I look forward to your favorably reply.

Sincerely yours,

Andy Liu

★

Dear Professor Lin:

My academic advisor, Dr. Liu, recommended that I contact you regarding the possibility of me receiving professional training in your research center on business management and related technologies during my upcoming summer vacation. Previous to my current study in the Master's degree program in the Institute of Financial Management at National Taiwan University, I acquired nearly five years of finance-related experience. Additionally, an undergraduate course in Business Evaluation Methods exposed me to the latest trends in finance and sparked my interest in researching related topics at the graduate level.

After completing our country's compulsory military service, I majored in Banking and Finance at Tamkang University and also worked part-time in the finance sector. Academic studies oriented me on many diverse topics, including financial statement analysis, commercial knowledge and industrial management. From my part-time work, I learned much about financial derivatives and assets trading. Additionally, I often spent my leisure time in preparing for examinations to acquire licenses that would qualify me work in Taiwan's finance sector. I plan to receive certification in the future that would allow me to practice in the finance profession in other countries.

As I am determined to pursue a research direction in business management research, the opportunity to receive training and practical working experience in your globally renowned organization would help instill in me a global perspective that would allow me to remain abreast of the latest trends in this dynamic field. I am also confident that I can acquire professional knowledge rapidly in order to keep pace with your colleagues.

Please feel contact my academic advisor, Dr. Liu, regarding my academic background and professional skills. I anxiously look forward to your thoughts regarding the above proposal, hoping that this opportunity would open future cooperative activities between our organizations.

Sincerely yours,

Ken Cheng

Dear Professor Porter,

I would like to apply for a self-supported practicum internship in your hospital. Allow me to introduce myself. After receiving a Bachelor's degree in Healthcare Management from Yuanpei University, I am pursuing a Master's degree in the Institute of Business Management at the same institution. Healthcare management has fascinated me since university, when I served in a practicum internship during my summer vacation of 2003 at Mackay Memorial Hospital in Hsinchu, Taiwan. Managerial practices in industry and a hospital markedly differ, largely owing to that the latter is a non-profit organization and is more closely related to an individual's welfare. After implementing the National Health Insurance (NHI) program in 1995, the Taiwanese government promulgated the Global Budget system in 2001, subsequently motivating hospitals to establish feasible targets in order to enhance their managerial and commercial practices to avoid a financial crisis both within individual hospitals and the overburdened NHI program.

Our country has much to learn from your country's well-established and comprehensive health insurance system. Receiving training and professional experiences in your hospital would allow to me to more fully understand how managerial approaches in Taiwanese and North American hospitals differ. Such an opportunity would allow me to apply in a practical context what I could learn from your hospital's experiences in clinical management.

Keen to remain knowledgeable of the latest hospital management approaches, I consider myself diligent and responsible whenever assigned a task. I hope to visit your hospital for two months next year, from July to August. My university will cover all expenses incurred during my stay.

Please carefully consider my application. I look forward to your favorable reply.

Yours truly,

Kua Shine Peng

Dear Dr. Smith,

As a graduate student in the Institute of Financial Management at National Taipei University, I am researching finance-related topics. My academic advisor suggested that I contact you regarding the possibility of receiving training in your company during my upcoming summer vacation. Such training would culminate in the development of a forecasting model for use in a finance report in Taiwan. My previous work in a company's finance department for more than two years involved accruing company expenses for an annual finance report. This work greatly oriented me on how results announced in a finance report can impact taxes levied on an enterprise's profits. While the Taiwanese government faces strained financial resources and constantly looks for additional revenue, the mandatory business income tax of 25% poses an additional burden on local and overseas investors. I am highly interested in learning practical measures to resolve this dilemma.

Given the opportunity to receive practical training in your company, I could increase my understanding of the latest finance-related research models and their applicability in the Asian Pacific region. I am confident that this training experience will yield many fruitful results, hopefully providing a valuable reference for the Taiwanese government in forecasting financial trends through the use of advanced software programs that would ultimately generate corporate revenues. Your company's leading role in analyzing and forecasting financial trends makes it a logical choice for me to acquire advanced professional skills in developing finance models for my graduate school research. Hoping to join your company from July to September during my upcoming summer vacation I will cover all expenses incurred during my stay.

I am especially interested in learning of new ways to offer investors various incentives. Thanks in advance for your careful consideration of the above proposal and look forward to your comments. I look forward to our future cooperation.

Sincerely yours,

Pin Huang

C. Exchanging Information／資訊交流信函

1.Proposing an information exchange relationship／資訊流通關係的提呈，請參考以下範例：

I'd like to propose an academic information exchange regarding current research trends in quality management in Taiwan and the United States.

Given your laboratory's innovative research in product marketing strategies, I would like to propose an information exchange relationship between our two organizations, hopefully beginning with my visit to your laboratory this upcoming July.

Since our goals closely match those of your organization, we would like to develop a partner relationship with you, for information exchange and for possible future collaboration.

Miss Hu recommended that I contact you regarding the possibility of an information exchange opportunity with your company with respect to potential technology transfer opportunities between our two organizations.

I look forward to exchanging laboratory data with your laboratory as part of a collaborative venture on designing long-term financial plans for the elderly.

2. Introducing one's professional experience or that of one's organization inrelation to the proposed information exchange relation / 為資訊流通而提供個人或機構經驗，請參考以下範例：

I belong to the Institute of Finance at National Taiwan University, a leading research center in this field in Taiwan.

In line with consumer demand, we continuously improve upon and streamline business management-related research methods.

As a governmental supported research organization, our Center maintains state-of-the-art instrumentation for clinical use.

While aspiring to become a globally leading research institute, our Institute hopes to participate in the activities of financial institutes worldwide.

I am currently involved in a research effort aimed at developing a process model by mining text-based information that will optimize management of customer relations in the business management sector.

第四個練習：

以下每句爲用來訓練有關資訊交流信函：爲資訊流通而提供個人或機構經驗。請把左邊及右邊的每一片語連結成一個完整句子。解答在本章最後。

1. While focusing on teaching, research and service, National Taiwan University has accumulated	a. my professor in a National Science Council-sponsored research project aimed at increasing the efficiency of customer service.
2. As a graduate student in the Institute of Technology Management at National Chiao Tung University, I am collaborating with	b. improve upon and streamline strategic management-related research methods and laboratory procedures.
3. The project that we are currently engaged in focuses on	c. I am involved in various stages of implementing National Science Council-sponsored research projects.
4. In line with consumer demand, we continuously	d. have yielded numerous benefits for the Taiwanese finance sector.
5. As a faculty member of the Department of Business Administration at Yuanpei University,	e. which are largely concerned with how to increase work productivity in the banking sector.
6. Enclosed please find a document that will hopefully give	f. to develop inexpensive yet high quality products for the burgeoning finance sector.
7. Let me briefly let you in on our ongoing projects,	g. the best research personnel and technologists in Asia, explaining its strong emphasis on teaching.
8. The work results generated so far in our project	h. branch office strives to offer a line of quality products to satisfy consumer demand, both locally and abroad.
9. Established in 1998, our company strives	i. providing relevant information that would facilitate a construction site manager’s decision making process.
10. As TFT-LCD technology continuously advances, our	j. further insight into the missions of the company which I belong.

3.Commending the organization or individual to exchange information with / 為資訊流通而讚揚某機構或某人，請參考以下範例：

The International Business Center of Stanford University is globally renowned for its pioneering research for practical applications.

Given your institution's state-of-the-art technology and advanced expertise in strategic management,

Given your headquarters' vast knowledge of consumer trends and technological developments in this area, we would like to exchange product information and manufacturing technology expertise with your R&D unit.

The achievements that your closely-knit staff has achieved are quite impressive.

Your recent research on Internet marketing has been extremely helpful in my graduate school research in the Institute of Business Management at Yuanpei University of Science and Technology, especially your unique strategies adopted to analyze the market potential for online commerce.

4.Outlining details of the information exchange relation / 資訊流通關係的細節詳述，請參考以下範例：

Through this relationship, we could freely exchange our views and approaches towards achieving the best available options for our customers.

As for the intern position offered at your organization on an annual basis, I could offer relevant data from Taiwan and the Pacific Asian region.

We are interested in what opportunities for technological cooperation are available between our two organizations in the area of developing forecasting models for the banking sector.

During our information exchange, we hope to collaborate with your laboratory in exploring current trends in the business management sector, hopefully enabling us to identify areas of mutual interest that would lead to a technology transfer or joint venture.

Our collaborative relation would hopefully allow us to share information on the latest advances in global finance.

第五個練習：

以下每句爲用來訓練有關資訊交流信函：資訊流通關係的細節詳述。請把左邊及右邊的每一片語連結成一個完整句子。解答在本章最後。

1. To initiate this information exchange relation, we suggest not only	a. to share pertinent data involving risk assessment, customer loan policies, database management and the transfer of business capital.
2. As our department actively engages in National Science Council-sponsored research in this area, we hope that exchanging relevant data between our two organizations will lead to	b. skills in evaluating the success factors of a strategic management project.
3. The information exchange between our two organizations will hopefully allow us	c. a joint venture aimed increasing the efficiency of the finance sector in Taiwan.
4. We hope to share our professional	d. of mutual interest, possibly leading to a technology transfer or joint venture.
5. I hope that our staff will have the opportunity to exchange	e. I could offer relevant data from Taiwan and the Southeast Asian region.

6. While you could provide valuable information on current trends in consumer behavior analysis,	f. that our personnel receive intern training on management planning in your institution, but also that your staff offer lectures and clinical demonstrations in Taiwan.
7. I hope that through our technological information exchange, we will be able to identify areas	g. data and experiences in the field of customer management.

5.Expressing optimism in the anticipated information exchange relation／表達對資訊流通關係的樂觀態度，請參考以下範例：

Familiar with the scope of activities in your marketing department, I am confident that our organization's similar methodologies and strategies will enable us to easily engage in a collaborative effort in the near future.

We believe that such a relation would be a mutually beneficial one.

We are anxious to develop long-lasting cooperative relationships with an institution such as yours.

I look forward to your thoughts regarding the possibility of such collaboration.

Such an exchange would enable us to acquire new perspectives towards the experimental process.

6.Initiating the information exchange relation／開始資訊流通關係，請參考以下範例：

Would you please send me introductory information as well as other pertinent literature that analyzes consumer behavior in the above areas?

Exchanging related information, e.g., methodologies, strategies, corporate missions and objectives, would appear to be the starting point for such cooperation, followed by visits to each other's marketing departments and, eventually, collaboration on a joint venture.

Please carefully consider my request and inform me of your decision at your earliest convenience.

I look forward to hearing your suggestions on how to initiate this information exchange relation, perhaps a visit to your marketing department to discuss relevant strategies and areas of mutual interest.

If you find such an arrangement agreeable, please notify me as to which dates are most convenient for you.

Sample Information Exchange Letters

★

Dear Professor Smith,

My academic advisor and your graduate school classmate, Professor Liu, suggested that I contact you to exchange data with you on investment sector-related research. As a master's degree student in the Institute of Business Administration at National Tsing Hua University in Taiwan, I received a Bachelor's degree in Information Management from the same institution.

Under the direction of Professor Liu, I am actively engaged in an investment-oriented research project involving a new area of interest in the insurance sector, Viatical Settlement. Given Taiwan's lack of experience in this area, Professor Liu recommended you as a valuable source given your expertise and your company's solid reputation in the American insurance sector. I am aware of your frequent discussions with Professor Liu on this matter and would like to establish contact with you so that we could exchange data that would be mutually beneficial to both of us.

My current project attempts to determine the feasibility of adopting Viatical Settlement in Taiwan's insurance sector, given how environmental factors in Taiwan differ from those in the United States. I have already accumulated data from local insurers in Taiwan through personal interviews on this topic in order to analyze the current business conditions in Taiwan and potential profits. Such results could also provide a valuable reference for you to more thoroughly understand Taiwan's investment climate and market trends. Given advances in global communication and the accelerated development of international investment, tremendous investment returns are owing to the diversity of available investment schemes, requiring strong analysis capabilities to thoroughly understand potential opportunities.

Thank you for your careful consideration of my above proposal. I look forward to establishing long-lasting ties with researchers in this field such as you for future collaborative activities. I hope to serve as an information source for your research or investment interests in Taiwan.

Sincerely yours,
Danny Wu

Dear Miss Ivy,

Thank you for your letter dated June the 10th, in which you offered valuable suggestions for our current situation. I would like to exchange information with you on efforts underway to reduce a high employee turnover rate through staff orientation and training.

Your organization has distinguished itself in human resources management, as evidenced by its strong corporate growth in Taiwan in recent years. Our company is developing a long-term educational training and communications program. This program ultimately strives to reduce the high employee turnover rate of technical personnel and to foster the professional skills of those personnel in a sustainable manner. Doing so entails devising a professional skills examination, holding educational training and analyzing the causes of a high employee turnover rate. Hopefully, results of this program will contribute to efforts aimed at maximizing the professional capabilities of its employees.

Given that a high employee turnover rate in our company has created prohibitively high personnel costs, the human resources department realizes the importance of fostering a nurturing environment for employees to maximize their potential. Our company has already initiated efforts to enhance communication channels between employees and management. Hopefully, successful implementation of this initiative will ultimately lower the employee turnover rate. Moreover, our company is anxious to share its experiences in this area with human resource departments in other organizations that face a similar predicament.

Please find enclosed relevant data of our current program for your reference. I look forward to hearing from you on how we could exchange relevant data that would be mutually beneficial to both of our organizations. If you agree with this information exchange opportunity, I will send you further program details next month. I would also appreciate any information that you feel would be beneficial. I would also like to arrange for a visit to your company to discuss the details of such cooperation. If possible, a tentative visiting schedule would be most appreciated. Thanks in advance for your careful consideration. I look forward to our future cooperation.

Sincerely yours,

Andy Liu

Dear Mr. Liu,

As a graduate student in the Institute of Business Management at National Chiao Tung University, I am researching finance-related topics involving business valuation practices in Taiwan. Your research organization has spent considerable time in addressing business mergers and acquisition-related matters, an area which our country is attempting to enhance its research capabilities. Therefore, I would like to propose that our two organizations exchange information and pertinent data on the latest trends in business mergers and acquisitions that would be mutually beneficial to both of our organizations.

Your research organization has significantly contributed to global valuation practices of business mergers and acquisitions, an area which Taiwan has much to learn. Taiwanese enterprises are generally small to medium-sized and lack global competitiveness despite its recent entry to the World Trade Organization. Global competitiveness of local enterprises can only be fully realized through business mergers or acquisitions.

The organization to which I belong is researching topics in the above area that are relevant to Taiwan's unique business climate. Much corporate data on various domestic enterprises, e.g., the electronics industry and the finance sector, have been accumulated, with a framework established to analyze such data. Establishing contacts with a globally renowned organization such as yours to exchange mutually beneficial data on the latest trends in this field would immensely boost our research efforts. Hopefully, such an information exchange opportunity would eventually lead to a technology transfer or joint venture in the near future.

We are excited about the possibility of exchanging pertinent data and the potential benefits of the long-term benefits of such a mutually beneficial relationship.
Sincerely yours,
Ken Cheng

Dear Mr. Smith,

Your country' s universal health insurance system is renowned for its successful implementation of the Global Budget System, an area which the National Health Insurance Bureau in Taiwan has become involved in recently given the unprecedented challenges and competition that the island' s medical organizations face. I would like to exchange information with your organization on research efforts underway in both of our countries to remain abreast of global trends in this area. Allow me to introduce myself. I received a Bachelor' s degree in Healthcare Management from Yuanpei University of Science and Technology in 2005. Several courses offered by the Department of Business Management enhanced my knowledge management skills. I am currently pursuing a Master' s degree in Business Management at the same institution. In addition to understanding how Taiwan implements the Global Budget System differently than your country does, I hope to devise strategies for local hospitals faced with bankruptcy as a result of implementing the Global Budget System.

I served in a practicum internship at Mackay Memorial Hospital in Hsinchu,

Taiwan during my summer vacation of 2003 as an undergraduate student. I began to realize how many hospitals that implement the Global Budget System potentially face bankruptcy when reducing reliance on governmental subsidies facing strong competition from other medical institutions. Your hospital has set a precedent for other medical organizations attempt to evaluate project expenses in order to avoid unnecessary risks.

Given the opportunity to establish an information exchange relation with your hospital, I hope that we could share pertinent data on efforts to implement the Global Budget System in our countries' health insurance systems. The results of such an information exchange could provide a valuable reference for hospitals in both of our countries to increase the efficiency of their operations.

To initiate this information exchange relation, I would like to visit your organization and discuss with you the details of our collaboration. Please let me know which dates for such a visit would be most convenient for you. Thanks in advance for your careful consideration of the above proposition. I look forward to hearing your favorable reply regarding this information exchange opportunity.

Sincerely yours,

Dear Mr. Smith,

I recently attended a financial management workshop, where we met briefly and exchanged business cards. As a graduate student in the Institute of Business Management at National Central University, I'd like to propose a technological information exchange regarding efforts to develop a model capable of identifying fraudulent financial reporting.

The company I work for, ABC, is a financial consultancy firm that assists customers in detecting fraudulent financial reporting. As our corporate goals closely match those of your organization, a partner relation between our two organizations would allow us to exchange related experiences, possibly leading to a collaborative effort in the near future.

Your organization has long distinguished itself in forecasting finance-related trends, as evidenced by your considerable efforts in training skilled professionals how to implement analytical and finance-related forecasting models. Would you please send me introductory information as well as any other relevant publications that describe your company's strategies, methodologies, achievements and future objectives? Specifically, we hope to share our professional experience with you in constructing analysis models capable of detecting fraudulent financial reporting, hopefully enabling us to identify areas of mutual interest that would lead to a technology transfer or joint venture in the near future.

We are optimistic that such a relation would be a mutually beneficial one, allowing us to develop long-lasting relations with a renowned institution such as yours. This opportunity would definitely allow us to identify areas of common interest that would benefit the directions that both of our organizations are taking. I look forward to this mutually beneficial, cooperative effort and hearing your ideas or suggestions regarding this information exchange opportunity. I would also like to arrange for a month technical visit to your organization this upcoming summer vacation. If at all possible, please notify me as to which dates are most convenient for you. A tentative visiting schedule would also be appreciated. Thanks in advance for your careful consideration. Please find enclosed our company introduction in DVD format, in which corporate missions, goals, strategies and methodologies are briefly introduced.

Sincerely yours,

Pin Huang

★

D. Making Technical Visits Overseas / 科技訪問信函

1. Requesting permission to make a technical visit / 科技訪問許可申請，請參考以下範例：

Although I strongly desire to receive an extensive period of training from your research group on the latest technological developments in this area, arranging such a large block of time is impossible given my hectic schedule at work. I would therefore like to arrange a three-day visit to your laboratory next month.

As I will be attending an international conference in New York with my colleague next month, we would like to visit the ABC Research Center in Los Angeles on February 15.

Given our interest in your technological developments in this field, we would like to visit your company to more thoroughly understand the progress that you have made.

Arranging for a five-day technical visit to your laboratory would appear to be the best way of beginning this discussion.

We would like to meet with members of your working group regarding details of our cooperative agreement, as well as tour your facilities.

第六個練習：

以下每句為用來訓練有關科技訪問信函：科技訪問許可申請。請把左邊及右邊的每一片語連結成一個完整句子。解答在本章最後。

1. Given the recent formation of this research group, I would like to arrange	a. discuss with you some of these topics in more detail.
2. Given my above interests, I would like to visit your research laboratory and	b. to tour your facilities and share our related experiences in product development.
3. I will be accompanied by	c. technical visit with the technical staff. in your Business Administration Department.
4. Given your corporation's committing to adopting the most advanced environmental protection technologies in your manufacturing processes,	d. for twenty days of technical instruction and consultation on the latest strategic management trends in your department.
5. I would like to arrange for a five day	e. as well as discuss the details of our cooperative agreement.
6. Given your company's expertise in this area, we would like to arrange for a technical visit next month	f. two environmental protection experts in this area.
7. As mentioned in our previous correspondence, we would like to visit your new computer testing facilities,	g. I would like to arrange for a three day technical visit to your product development division to learn of the operational aspects and theoretical applications that you adopt in production.

2.Stating the purpose of the technical visit／闡述科技訪問的目地，請參考以下範例：

We would like observe a clinical demonstration of the latest computer software that can reduce examination time as much as possible.

The ability to integrate the advances of your unique research program with the unique aspects of image fusion would significantly contribute to our ongoing efforts.

In addition to touring your facilities, we hope to discuss with you potential areas of mutual interest, as outlined below.

The opportunity to consult with you in person will hopefully yield mutually beneficial results for both of our laboratories.

I am eager to learn of the potential implications of this new technology for ongoing research efforts already underway island wide.

3.Introducing one's qualifications or one's organizational experiences／介紹某人的能力或機構經驗，請參考以下範例：

Our company's Product Development Division is committed to adhering to the Taiwanese government's aggressive policy towards promoting the use of environmentally friendly practices in manufacturing.

Let me brief you on our ongoing projects.

Technological advances in quantitative analysis and computer software have greatly facilitated our laboratory's efforts in researching marketing approaches adopted in the global network and their implications for marketing strategies aimed at the Greater China sector.

My graduate school research attempted to elucidate the actual mechanisms that are present during a technology transfer.

With aspirations of becoming a "Green Silicon Island", Taiwan has actively encouraged its enterprises in recent years with numerous incentives to be more socially responsible in its manufacturing practices.

4.Commending the organization to be visited／對訪問機構的讚揚，請參考以下範例：

I found the results from your recent article to be extremely helpful to my own research direction.

As is well known, your company pioneered the use of strategic management in distributing daily goods and appliances.

As a global leader in developing medical instrumentation, your laboratory definitely has much to contribute to our country's efforts to elevate its research capacity in this area.

In this area, your laboratory has acquired much expertise and relevant data, explaining why I am eagerly looking forward to this upcoming technical visit.

Your products conform to the most stringent environmental standards by using ingredients that are phosphate-free and environmentally friendly.

5.Outlining details of the technical visit／詳述科技訪問的細節，請參考以下範例：

I am especially interested in learning of your successful cases of treatment to reduce the likelihood of a stroke from occurring.

To give you time to prepare in advance for our discussion, I would like to briefly summarize some of the points that we hope to cover during our time together.

I would like to pose the following questions before our discussion on November 12 so that you will have sufficient time to prepare your responses:

We are especially interested in discussing this concern with you, particularly the current technological status of investigating transgenic animals.

I am open to any suggestions you might have regarding the itinerary or any materials you would like me to prepare prior to my visit.

6.Requesting confirmation for approval of the technical visit／科技訪問核准的確認申請，請參考以下範例：

If this visit is agreeable with you, please notify me as to which dates are convenient for you.

To arrange for the upcoming research visit in order to consult with you on the above topics, I need to set up an itinerary and time schedule to report to my superiors.

Quite some time has passed since I originally proposed the idea of our technical visit to your country without any response from you so far.

Prior to our visit, we would appreciate any introductory information that you could pass on to us regarding transportation arrangements and a list of hotels near our meeting place.

A suggested itinerary would also be most appreciated, including our discussion time, tour of research facilities and accommodations.

Sample Technical Application Letters

Dear Mr. Johnson,

As an employee of Chunghwa Telecom, the largest telecommunications provider in Taiwan, I will be attending a communications technology convention in Silicon Valley and hope to have the opportunity to visit your company during that period. Such a visit would tremendously boost our company's product developments efforts given your well-established reputation in Silicon Valley and state-of-the-art facilities. I hope to acquire the latest information on recent trends in product development that would possibly lead to a technology transfer or joint venture in the near future.

Having just received a Master's degree in Business Administration from National Sun Yat Sen University in Taiwan, I obtained a Bachelor's degree in Information Management from Nanhua University previously. This solid academic background exposed me to the latest knowledge skills in Internet-based technologies from a management perspective, enabling me to conduct graduate level research independently. Confident of my own abilities, I am more than willing to share with you a summary of our own product development efforts and managerial approaches from a Taiwanese perspective.

As a widely recognized telecommunications provider globally, your company plays a leading role in this field of research, as evidenced by your continuous development of Internet-based technologies and instrumentation to enhance global telecommunications. The increasing number of telecommunication technologies available have significantly elevated living standards and made daily life more convenient. The opportunity to visit your company with provide me with exposure to advanced technologies and state-of-the-art equipment that your research department has spent considerable time in developing. This visit would most definitely provide a valuable reference for our company's research endeavors, which will hopefully prove mutually beneficial for both our organizations through a possible technology transfer or joint venture in the near future.

As I will be staying in a hotel in Silicon Valley during my stay in the United States, you can communicate with me directly through e-mail. Thanks in advance for arranging this visit. I look forward to our future cooperation.

Sincerely yours,
Danny Wu

Dear Mr. Chen,

Thank you for your prompt response via e-mail and dated October 20, 2005, regarding my query on investment banking practices and business merger services of Credit Suisse First Boston. I am eager to visit your research center in New York, which would not only orient me the recently emerging financial engineering field, but also provide a valuable reference for my current research project on reform within Taiwan's finance sector.

Our company division is actively engaged in a project on reform within Taiwan's finance sector by continually upgrading the island's competitiveness and innovativeness through adoption of state-of-the-art finance engineering technologies. Our company has already accumulated a considerable amount of finance sector-related data for analysis and research purposes. Financial models developed based on that data accurately reflect Taiwan's unique business climate. As Taiwanese researcher hope to remain abreast of the latest developments in this field, the opportunity to visit your research center would be initial step for our two companies to engage in a collaborative effort in the near future.

Selecting your research center for a visit was a logical choice given your global renown for handling merger and acquisition cases. During my visit, I hope to consult with Mr. Chen and his working group to learn of your company's unique operational approach. I look forward to any suggestions regarding this visit and materials that you would like us to prepare beforehand.

Accompanied by two of my colleagues, I hope to arrange for a visit to your research center and other relevant departments this upcoming December. If this visit is agreeable with you, please notify me which dates are convenient for you. A suggested itinerary would also be most appreciated, including discussion time, plant tour schedule and accommodations. I look forward to your reply. I am confident that this visit will open doors for further cooperation between our two organizations.

Thanks for your assistance.

Sincerely yours,

Ken Cheng

★

Dear Mr. Smith,

I would like to arrange an interview with you on your company's current management direction as part of my graduate school research at National Tsing Hua University. Hopefully, the results of these interviews will provide a valuable reference for you as well. The interview format is clearly defined, with topics to be covered provided to you in advance. I will closely work with your company to ensure that the interview yields optimum results.

This interview will largely focus on your company's current operations, future management directions and brand image appeal. Results of this interview will also provide a valuable reference for company shareholders hoping to understand your current management approach, thus facilitating your efforts to more closely interact with investors. My colleague, Ms. Grace, will be responsible for handling the interview. She is well experienced in this area, as evidenced by her congenial and professional attitude towards respecting the technical expertise of each interviewee. Each interview will be conducted according to the conventionally adopted 6 S standard principles of skillfullness, smiling and congenial approach, speed, specialization, standardized procedure and spirit of cooperativeness. Moreover, interviews are held using state-of-the-art sound recording equipment to filter out outside interference. Many enterprises have responded favorably to our interview techniques and the subsequent results generated.

The three hours allotted for the interview will hopefully allow us to gain a clear picture of your company's management and technology investment directions. Interview results will be sent to your company in a detailed report form shortly after carefully analyzing the results and expressing them in a concise manner. If the above is satisfactory, please let us know which time and location are the most convenient for you.

Sincerely yours,

Andy Lin

★

Dear Mr. Smith,

Taiwan's National Health Insurance Bureau has recently implemented the Global Budget system to ensure fiscal responsibility among the island's healthcare institutions, posing a major challenge to strictly control operational costs while maintaining quality medical care. Many hospitals adopting the Global Budget system have lost considerable revenues and even face potential bankruptcy.

Taiwan has much to learn from your country's comprehensive health insurance system that has successfully risen to the challenges of the Global Budget system.

As I am completing my master's degree in Business Administration at National Taipei University, my academic advisor recommended that I closely examine your hospital's management initiatives in this area.

Given your hospital's innovative management approaches in providing quality medical care, I would like to arrange for a visit to your hospital to understand how it copes with strained financial resources while implementing the Global Budget System.

During my upcoming visit, I hope to discuss with you the merits and limitations of three management approaches that I am interested in implementing the following:

1. Establishing a polyclinic on the hospital premises, commonly referred to as a clinic in the front door. Doing so enables hospitals to receive more patients and generate additional revenues from examination fees. As an independent entity under the control of the hospital, a polyclinic draws its resources from the hospital as well, including medical personnel and instrumentation;
2. Establishing a polyclinic by forming a strategic alliance with another clinic or offering non-NHI subsidized services to increase hospital revenues; and
3. Adopting multi-disciplinary management practices in hospital operations to remain competitive in the intensely competitive medical care sector. Notable examples include establishing clinical services that do not fall under the NHI scheme, such as cosmetic services, physical fitness and weight reduction training and health examination services.

If convenient, I hope to consult with you in person on the above topics. Let me know when such a visit would be most convenient. Please carefully consider my request. I anxiously look forward to your favorable reply.

Sincerely yours,

Kua Shin Peng

★

Dear Mr. Smith,

Thank you for allowing us to visit to your company on November the 21st.? As mentioned in my earlier correspondence, I am attempting to upgrade the precision of a forecasting model to ensure its applicability in Taiwan's finance sector. However, according to preliminary results, our model still does not provide a reliable and effective means of reducing the probability of error. As a global leader in developing financial forecasting models, your company could offer much valuable advice in our efforts to perfect our forecasting capabilities in the Asian Pacific region.

Our ongoing project is to detect fraudulent behavior in financial statements. Studies devoted exclusively to detecting fraudulent behavior in the finance sector are relatively scarce. Whereas related studies largely adopt a logistic prediction model, your company has taken a leading role in analyzing and forecasting financial trends.

My previous work in a company's finance department for more than two years focused mainly on accruing company expenses for an annual finance report. This work greatly oriented me on how results announced in a finance report can impact taxes levied on an enterprise's profits.

ABC Company is recognized as the global leader in analyzing and forecasting financial trends. Analytical procedures are widely regarded as a highly effective means of detecting fraudulent behavior. Touring your research facilities would therefore be an educational and enriching experience. I am especially interested in learning of your advanced analysis and forecasting software programs.

Hopefully, the opportunity to visit your company and more thoroughly understand our organizational needs could lead to a joint collaboration in the near future. As I must finalize my trip itinerary before November 10, 2005, your speedy reply would be most appreciated.

Sincerely yours,

Pin Huang

★

E. Inviting Speakers and Consultants／演講者邀請信函

1. Inviting a speaker／邀請演講者，請參考以下範例：

Given your renowned research and contributions in the field of strategic management, we would like to formally invite you to serve as an Invited Speaker at the upcoming Global Management Symposium.

Having extensively read upon your published research on supply chain management, I find these results most helpful to my own research direction.

We also hope that you could chair a roundtable discussion on a related topic.

Given your distinguished research on the Helicobacter pylori full genome, I would like to invite you to speak at the upcoming Ninth Symposium on Recent Advances in Cellular and Molecular Biology.

Given your organization's extensive experience in implementing long-term care programs, we hope that you could recommend a consultant that could instruct our hospital staff on the following topics:

2. Describing the purpose of the lecture／描述演講目地請參考以下範例：

Your lecture will hopefully include a summary of recent developmental trends in medical science, relevant governmental policy and current status of medical education.

Your lecture will hopefully include a summary of recent developmental trends in hospital management, relevant governmental policy and current status of medical education.

The theme of this seminar will coincide with efforts to develop product technologies in this area.

Despite the above aspirations, our hospital is lacking in many of the above areas, explaining the lack of confidence among patients and their relatives in the quality of medical care offered in Taiwanese hospitals.

With more than 200 practitioners and academics anticipated to attend the symposium, participants hope to learn of the latest medical imagery trends in this area.

The seminar will focus on recent advances in management science.

3. Explaining details of the lecture／詳述演講的細節，請參考以下範例：

Each lecture will last two hours, with the consultant reimbursed with a speech honorarium.

I am pleased to announce that a seminar on DNA mutation, methylation, repair and recombination will be held at Yuanpei University of Science and Technology in July of 2005.

As February is normally a busy month in preparation for the Chinese New Years holiday, I would suggest rescheduling the symposium for March the 22nd.

Perhaps you could also recommend a speaker who could discuss the features of Internet-based technology applications.

My colleagues will contact you regarding details of the symposium and travel arrangements.

As the Symposium will coincide with the Christmas holidays, we will also arrange a special Christmas dinner and related activities.

I need to know which dates are the most convenient for you to give this lecture.

4.Describing transportation and accommodation details／詳述交通及住宿的細節，請參考以下範例：

We will provide roundtrip airfare and accommodations during your visit.

Besides providing accommodations and transportations arrangements during your stay, we would also like to show you some of Taiwan’s cultural and historical places of interest.

All accommodation and transportation arrangements will be made in advance.

I advise that you stay here for an additional week or so to see some of Taiwan’s scenic spots and places of historical significance.

In addition to the symposium, we will arrange several trips to enjoy some of the sights that Taiwan has to offer.

第七個練習：

以下每句爲用來訓練有關演講者邀請信函：詳述交通及住宿的細節。請把左邊及右邊的每一片語連結成一個完整句子。解答在本章最後。

1. As for compensation of this event, please pay	a. for a single room averages between DM$ 130 to $US 250 daily.
2. Please allot extra time besides your involvement in the conference	b. I need your formal name, current address and passport number.
3. To reserve your air ticket and hotel accommodations at the Taipei Spring Leisure Resort,	c. a roundtrip train ticket and hotel accommodations in Hsinchu for the evening of May the 15th.
4. In addition to an honorarium for your lecture, the organizing committee will also provide	d. that you will be able to tour some of Taiwan's natural scenery.
5. Incidentally, hotel accommodations in Taipei	e. in advance for the roundtrip airfare ticket, and other incidental expenses.
6. Please inform us of your arrival time to the Hsinchu	f. reserve a room for you and your wife.
7. I hope that you can allot extra time to see some	g. of the cultural wonders that Taiwan offers.
8. Once your schedule is confirmed, I will	h. to enjoy some of the island's cultural and natural wonders.
9. I advise you to lengthen your visit until March 26 so	i. train station so that we can meet you.

5.Instructing the speaker what to prepare before the visit / 演講者行前必知指導，請參考以下範例：

Please send us your lecture titles and related handouts before January 10, 2005 so that we will have sufficient time to make copies and appropriate arrangements.

Please e-mail us your curriculum vitae and course materials before December 24, 2004 so that we will have sufficient time for translation and printing.

I need your curriculum vitae, with contents to include the following name, date of birth, place of birth, nationality, address, e-mail address, telephone number, marital status, academic qualifications, professional experience, scientific achievements, current scientific activities, other science-related activities and selected publications.

Please send us the text of your talk before December 5.

Let us know if you have any special requirements prior to your presentation or on the day of the seminar.

Sample Invitation Letters

Dear Professor Smith:

Given your renowned research and contributions in the field of computer science, we would like to formally invite you to serve as an invited speaker at an upcoming information technology symposium in Taiwan. Your lecture will hopefully include a summary of advanced computer hardware and software, as well as your current research directions - all of which will provide a valuable reference for domestic efforts to upgrade information technology capabilities. The symposium intends not only to orient local industrial professionals and academics on the latest developments in computer technology, but also to provide a forum of discussion that would facilitate collaborative opportunities, enabling the local information technology sector to effective respond to consumer demands.

The symposium will be held at Industrial Technology Research Institute (ITRI) in Hsinchu on December 1, 2005. Let us know in advance any special equipment or preparations that you require for your seminar presentation. If you have other commitments during this time, could you recommend another speaker qualified to analyze future consumer demand in the information technology sector.

ITRI will provide accommodations and transportation during your stay, as well as a speech honorarium for your lecture. In addition to the symposium, our colleagues will take you sightseeing some of Taiwan's scenic spots so that you can enjoy some of the island's natural wonders.

Please send us your title and lecture handout before November 15, 2005. Please let us know of any last minute preparations to be made before November the 25th. Once plans are finalized, we will forward to you the airplane ticket along with all flight details so that we can meet you upon your arrival to Taoyuan International Airport.

While drawing computer engineers from around Taiwan, this annual symposium will profoundly impact efforts to foster a lively discussion on the latest advances in information technology. Your contribution will definitely benefit all of the attendees. We look forward to your favorable reply.

Sincerely yours,
Danny Wu

Dear Professor Wu,

Given your expertise in marketing and human resource management, we would like you to lecture on human resource management trends in community development at academic symposium to be held at National Chiao Tung University on December the 5th , 2008. Your presence would greatly add to the success of this event. Details regarding the lecture and a roundtable discussion thereafter are as follows. The lecture will begin at 9 am and last approximately two hours. An interim break will last twenty minutes, with refreshments served. Following a 30 minute question and answer session as well as a recess for lunch, , the roundtable discussion will last from 1-4 pm.

The organizing committee will make all transportation arrangements, , as well as provide accommodations during your stay. Tentatively, your air flight will arrive at Chiang Kai-Shek International Airport in Taoyuan, with a shuttle service transporting you to the Ambassador Hotel in Hsinchu. Besides the symposium, our colleagues will arrange a sightseeing tour of Hsinchu City, which will last the greater part of a day.

I hope that you find the above arrangements acceptable. Please feel free to contact me via telephone, e-mail or post if you have any comments or suggestions. Once you have confirmed that you will be able to participate, I will send you the symposium schedule and your personal itinerary.

The Graduate Institute of Business Management will host the roundtable discussion. Your knowledge expertise will prove invaluable for all attending staff and graduate students. Thanks in advance for your careful consideration of our request. We look forward to your favorable reply.

Sincerely yours,
Wei-Hsiang Liu

★

Dear Mr. Stevens:

Given our strong interest in the research direction that you are taking and your abundant research expertise, the Institute of Finance Management at National Tsing Hua University would like to invite you to serve as an invited speaker at a seminar to be held in Hsinchu.. Hopefully, you could share with us with your research experiences that would contribute to our own efforts. We are especially eager to learn of the advanced research methodologies that you adopt. . We are currently limited in the inefficient methods utilized in our research.

If you are available, I will send you a street map of our school' s location. You will also be reimbursed for transportation expenses, as well as provided with a speech honorarium for the lecture. A contact person will be in touch if you can attend, as well as hotel reservations made if necessary.

If you are free, please stay after the lecture and accompany us on a sightseeing tour of some of Hsinchu' s historical attractions in which you can also enjoy the local cuisine. Also, let us know of any dietary restrictions that you have for meals during your stay.Before your arrival, please send us your lecture topic and related handouts so that we can make copies on timeYour lecture will definitely benefit our researcher. Looking forward to your favorable reply.

Sincerely yours,

★

Dear Professor Wu,

Given your significant contributions in the health promotion field, we would like to formally invite you to serve as an invited speaker at a symposium to be held on February 28, 2006 at Yuanpei University of Science and Technology (YUST) in Hsinchu, Taiwan. The symposium will begin at 9:30 am, with each lecture lasting roughly one hour, with an additional ten minute for a question and answer session. Hopefully, your lecture will focus on the practical implications of health promotion and the latest trends, along with recommendations on how to promulgate this concept effectively in Taiwan. While pointing out areas of particular concern in promulgation of such a strategy, you could also share your own personal experiences in this area.

We will provide roundtrip airfare and hotel accommodations during you stay in Taiwan, as well as a speech honorarium for your lecture. I hope that you can stay in Taiwan for a week of sightseeing the island's natural and cultural wonders, as well as enjoying local cuisine. The travel program will be drawn up with your interests in mind.

Once your plans are finalized, please let us know your flight details so that we can meet you at the airport on February 27, 2006. Then, on the morning of February the 28th, a taxi will transport you from your hotel in Taipei to YUST in Hsinchu. All of your accommodation and transportation arrangements have been carefully planned.

Please send us your lecture title and handouts before January 31, 2006 to allow sufficient time for copying. Let me know if you have any concerns regarding the above arrangement. Your contribution to this symposium will definitely benefit all of participants. We look forward to your favorable reply.
Sincerely yours,

★

Dear Mr. Smith,

Owing to your eminent research and significant contributions in the financial analysis field, we would like to formally invite you to serve as an invited speaker at an upcoming finance symposium to be held in the International Conference Hall at Yuanpei University of Science and Technology in Hsinchu, Taiwan, on March 1, 2006. Hopefully, your lecture will cover the latest trends in finance and accounting practices..

Given your organization's global reputation in financial analysis use of advanced software programs in this area, we hope that your lecture could address some of the following topics:

1. Analyzing the corporate health of an enterprise for investment purposes
2. Forecasting the anticipated annual revenues of an enterprise
3. Maximizing the profits of current investors
4. Maximizing the value of stock shares

Incorporating the above topics into your lecture would greatly benefit our corporate staff.

Please pay in advance for your roundtrip airfare and hold onto receipts for other incidental expenses so that we can reimburse you before you leave Taiwan. Besides the symposium, please allot extra time to sightsee some of the island's natural and cultural wonders.

If the above arrangement is acceptable, please send your curriculum vitae, including your name, date of birth, nationality, academic qualifications, professional experiences, achievements in the finance sector, current finance-related activities and selected publications. Also, please send us your lecture title and related handouts before February 5, 2006. I will be in touch with you shortly regarding further details of the symposium and travel arrangements. Your contribution to the symposium would definitely benefit all of the participants. We look forward to your favorable reply.

Sincerely yours,
Chih-Pin, Huang

F. Arranging Travel Itineraries / 旅行安排信函

1.Acknowledging the upcoming visit / 旅行安排信函，請參考以下範例：

Thank you for your response regarding Mr. Smith's visit to our Taiwan branch office.

Thanks for your recent reply that Mr. Robbins has agreed to participate in our academic discussion on intellectual capital-related topics in Taiwan.

Thank you for agreeing to serve as an invited speaker at the upcoming seminar to be held at Yuanpei University of Science and Technology, November 22-30, 2004.

Your participation in the upcoming symposium will definitely benefit all of the attendees.

Thanks for your positive reply to our invitation of Dr. Kanbinsky as a guest speaker at the PIDA technical seminar on current trends in laser technology.

2. Listing details of the travel itinerary / 行程細節列表，請參考以下範例：

Meet with R&D staff in the Taiwan branch office and discuss Mr. Smith's schedule.

Tour the northern coast of Taiwan and experience the island's art of drinking tea.

Deliver an introductory lecture on cyclotron and synchrotron.

Discuss areas of cooperation in developing innovative product technologies.

Continue with discussion on areas of cooperation between our two organizations.

Tour the campus of National Tsing Hua and National Chiao Tung Universities.

Spend free time sightseeing in Taipei.

3. Welcoming suggestions about the proposed itinerary / 對行程細節列表的建議，請參考以下範例：

The above itinerary and consulting topics are only tentative.

Any comments or suggestions regarding the aforementioned proposal would be highly appreciated.

Let us know if you have any suggestions regarding the above schedule.

The agenda and consulting topics can be changed, if necessary.

If you have any valuable suggestions regarding the above agenda and lecture topics, please do not hesitate to contact me.

Sample Travel Itinerary Letters

Dear Mr. Wu

Thank you for agreeing to to serve as an Invited Speaker at the upcoming symposium on Financial Innovation and Developmental Trends, which will be held in the International Conference Hall at Yuanpei Institute of Science and Technology (YUST) in Hsinchu, Taiwan on December 15, 2005. Given your eminent research in business mergers, acquisition technologies and financial analysis, we hope that you could also chair a roundtable discussion on the following topics:

1. Evaluating how Japanese governmental authorities facilitate financial reform
2. Assessing how business mergers and acquisition technologies assist in the

financial reform of Taiwanese enterprises3. Determining the merits and limitations of

financial reform of Taiwanese enterprises

The following is a tentative schedule for your visit:

December 13, 2008: Arrive in Taiwan at Taoyuan International Airport, followed by transit to Ambassador Hotel in Hsinchu

December 14, 2008: Rehearse for the symposium in the Conference Hall at YUST

December 15, 2008: Lecture on how to assess financial reform of Taiwanese enterprises and chair a roundtable discussion following the lecture

December 16, 2008: Sightsee Hsinchu's cultural attractions and enjoy local cuisine

December 17, 2008: Return to the United States

Before your arrival, please send us your lecture title, handout and curriculum vitae.

Sincerely yours,

Becky Li

★

Dear Dr. Martin:

Thank you for the warm hospitality that you extended during our visit to your laboratory, as well as the informative project results that you shared with us. A summary of those results was orally presented to our colleagues upon our return to

Taiwan. They were quite impressed with the progress that your laboratory has made so far. Therefore, we would like formally invite you to serve as an invited speaker at a conference that coincides with our hospital's 50th anniversary celebrations. The event will be held in Taipei on April 23, 2004, with the International Conference Hall at National Taiwan University serving as the conference venue.

The following is a tentative schedule for your visit:

September 1, 2008: Arrive at Chiang Kai-Shek International Airport in Taoyuan, with a bus shuttle service to your accommodations provided.

September 2, 2008: Consult with our hospital administrators on areas of improvement and potential areas of collaboration between our two organizations.

September 3, 2008: Deliver first lecture.

September 4, 2008: Deliver second lecture.

September 5, 2008: Visit Chiang Kai-Shek Memorial Hall, National Palace Museum and the Taipei Financial Center.

September 6, 2008: Sightsee local cultural attractions in Taipei.

Please pay in advance for your roundtrip airfare and hold onto receipts for other incidental expenses so that we can reimburse you before you leave Taiwan.

If you have suggestions regarding the above itinerary, please do not hesitate to contact me. If you are able to attend this special event, please forward your lecture topic and related handouts to Dr. Ding before July 2, 2005 to allow sufficient time for printing.

Your contribution to this conference would definitely benefit all of the participants.

Sincerely yours,

Thomas Lin

G.Requesting Information / 資訊請求信函

1.Explaining why contact was made to request information / 解釋請求資訊的目地，請參考以下範例：

While I was speaking with your products manager Mr. Li over the phone regarding the purchase of your company's electronics equipment, he suggested that I directly contact you to obtain your company's latest products catalog.

After I talked to Mr. Yang over the phone regarding your line of TFT- LCD products, he suggested that I contact you.

Thank you for sending us your latest company catalogue and product samples. Your product offerings are in line with our organizational needs.

Given your pioneering work on the decision theory, Analytic Hierarchy Process (AHP) and the Analytic Network Process (ANP), I have read many of your published articles and am well aware of your more than twelve books on these topics.

Your recent presentation on the database system containing long term care-related information, as developed by your company, was most informative.

2.Describing organizational need for information／資訊請求信函，請參考以下範例：

Our company is currently involved in a project to develop slide systems that involves the use of TFT- LCD panels.

The project we are participating in is part of a government-sponsored collaborative effort with U.P.O. Company, aimed at contributing to our country's information technology infrastructure.

Although such data for Taiwanese hospitals is scarce, I am actively engaged in researching related topics.

We are interested not only in mining the text information with content analysis method, but also in integrating the information to construct the data warehouse for the same analytical purposes.

In addition to our institution, several other universities have expressed to us their interest in incorporating your system into their daily operations and would like to acquire further information on purchasing, installation and customer service.

3.Directly requesting information／資訊請求信函，請參考以下範例：

Does your company have a sales representative in Taiwan? If so, please ask him/her to contact me.

Having gone thoroughly read your article, I was wondering whether you would send me pertinent literature or materials on the following areas:

In light of our budget constraints, we would appreciate it if you would reconsider your original quotation.

I hope that you will be able to provide us with some of the above detailed information.

Could you provide us with information from your statistical database on health care in the United States?

Sample Information Request Letters

Dear Dr. Smith,

Allow me to introduce myself. Having received a Bachelor' s degree in Business Management from National Central University recently, I am pursuing a Master' s degree in the Institute of Business Management at the same institution. As I am researching healthcare management and health promotion management-related topics, my master' s thesis will focus mainly on assessing the status of health promotion in local hospitals. As is well recognized, health promotion is of priority concern among hospital administrators worldwide. My academic advisor recently informed me of your efforts in popularizing a health promotion strategy among hospitals in your country.

Given your experience of more than a decade in health promotion, I have been closely monitoring the development of the health promotion project that you are spearheading. Health promotion heavily emphasizes assessing healthcare organizations, patients, staff and the surrounding community. To ensure success, a health promotion strategy must be implemented in line with consumer demand. Our current project involves collaborating with National Taiwan University Hospital to construct a health promotion model that is appropriate for Taiwan' s current needs and in line with mandates of the World Health Organization.

Having accumulated extensive knowledge expertise of health promotion research in a clinical setting, you are a valuable source of information in this growing area of research. I would like to draw upon your expertise by learning more of the successful case studies you have been involved with in your innovative research I would appreciate any pertinent data that you could pass along on how to conduct such research in a clinical setting. Specifically, I am interested in current health promotion strategies, organizations that have successfully implemented such strategies and profiles of patients, staff and communities involved in previous case studies. If convenient, could you send me a list of hospitals involved in your health promotion projects. Thank you in advance for your careful consideration of my research interests in this area. I would be more than happy to provide you with pertinent data regarding efforts underway in Taiwan. I look forward to the opportunity to cooperate with you.

Sincerely yours,
Mei-Hsuan Kuo

Dear Mr. Stevens:

It was a pleasure meeting you during your recent visit to Taiwan. I have closely followed your research for quite some time, as you are widely revered in the marketing field for your innovative efforts. Allow me to introduce myself. I am pursuing a Master's degree in Business Management at Yuanpei University of Science and Technology, where I also received a Bachelor's degree in Food and Beverage Management.

Given that our research directions are quite similar, I have widely read upon your previous research and am deeply impressed with the way you articulate your research findings in a concise and easy to comprehend manner. Your published articles contain many innovative concepts that have been most helpful in my graduate school research. With your specialization in marketing research, my academic advisor often encourages me to remain abreast of your latest research findings.

Despite my academic fundamentals in marketing, I lack professional experiences, explaining why I am writing to you to seek guidance on understanding some of the more practical implications of this research discipline. In addition to refining my knowledge skills, graduate school has exposed me to a wide array of integrated marketing practices. Closely following your research has enabled me not only to remain attune of the latest marketing trends, but also to learn how a well- structured and concise research article should be written.

As I constantly strive to enhance my specialized knowledge of marketing, I would appreciate any comments or related references that you could pass on for my current research direction. I am especially interested in a reading list that you assign to researchers in your laboratory. I look forward to hearing from you at your earliest convenience.

Sincerely yours,
Ei-Hua Wu

Dear Dr. Kandinsky:

Allow me to introduce myself. After receiving a Bachelor's degree in Healthcare Management from Yuanpei University, I am now pursuing a Master's degree in the Institute of Business Management at the same institution. Healthcare management has fascinated me since university, when I served in a practicum internship during my summer vacation of 2003 at Mackay Memorial Hospital in

Hsinchu, Taiwan.

Having read many of your published articles, I am well aware of your significant contributions towards creating a fiscal balance in the payment system of the national health insurance (NHI) schemes. The NHI Bureau in Taiwan has recently implemented a Global Budget System, posing a major challenge for medical organizations to strictly control operational costs while maintaining quality medical care. Many hospitals adopting the Global Budget system have lost considerable revenues and even face potential bankruptcy. Our country has much to learn from your country's well-established and comprehensive health insurance system. I was wondering whether you could send me literature or materials in the following areas:

A brief introduction of approaches towards implementing the Global Budget System in your country

A summary of the current status of various countries in implementing the Global Budget System

A brief introduction of how the Global Budget System has influenced the NHI scheme in your country

A brief synopsis of future trends in the Global Budget System

To maintain their long term competitiveness and survival, hospitals can adopt any of the following three schemes: establishing a polyclinic, forming a strategic alliance with another clinic or offering non-NHI subsidized services to increase hospital revenues. My graduate school research focuses on developing an ANP or AHP-based model to assist hospital administrators to make the most appropriate decision. I am especially interested in how various countries implement the Global Budget System.

I hope that you will be able to provide the above information. Also, if you find my area of research compatible with yours, we could perhaps collaborate in some way in the near future. I would be more than happy to provide you with pertinent data regarding NHI trends or efforts to implement the Global Budget System in Taiwan. Thank you in advance for your kind assistance. I look forward to our future collaboration.

Sincerely yours,
Kua-Shin Peng

★

Dear Mr. Lin:

While listening intently to your recent lecture at National Taiwan University, I was most intrigued with your research on investment banking practices and business mergers, areas that I hope to gain further expertise. I am especially interested in adopting your research framework and data from your pioneering work on decision theory. Among several of your published articles that I found quite helpful in my own research, I especially enjoyed your articles "Valuation of Business Methods" and "Investment Banking and Business Merger of Theory and Practices"

As is well known, business merger methods have been extensively adopted not only in business and industry, but also in the valuation of company stock prices and in decisions to raise or lower the amount of capital invested. Such decisions profoundly impact decisions involving potential mergers and the determination of surplus value in a company for future investment purposes.

Specifically, I am interested in the following:

1. What decisions are involved in valuation methods aimed at assessing commerce and industry?
2. What variables are included in the valuation model?
3. How are valuation methods related to each other?

The Taiwanese government has aggressively advocated reform of the financial sector in recent years to continually upgrading the island's global competitiveness and innovativeness through adoption of advanced finance-related models and approaches. My master's degree thesis is attempting to evaluate the success of such reforms.

I would be most appreciative if you could pass on to me pertinent data regarding the role of investment banking in a business merger or valuation in the United States. I would be delighted to send you any pertinent data from Taiwan so that our exchange could be a mutually beneficial one.

Thanks for your kind assistance.

Sincerely yours,
Ken Cheng

Dear Professor Buckley:

As a graduate student in the Institute of Business Administration at National Central University, I immensely enjoyed reading the article that you published recently in the European Journal of Operational Research, entitled "Fuzzy hierarchical analysis revisited". This topic is also the focus of my master's thesis research. Your insightful article helped me resolve some obstacles encountered during calculation of my research data while using the Fuzzy AHP approach. Despite an extensive online search on the use of the approach in the particular context of my research, I have been unsuccessful in obtaining pertinent information. This explains why I am writing to you directly in order to locate more relevant information sources.

I am currently developing a Fuzzy AHP-based method to evaluate the site selection of adult and juvenile correction facilities in Taiwan. Our graduate school instructors have oriented us on how to use various statistical methods, including AHP and the grey forecasting methods. AHP appears to be the most appropriate for my research topic, as I have already found several similar articles from an Internet search. Given your expertise in using Fuzzy AHP and my current research in this area, I would appreciate any assistance that you could offer, such pertinent references or information on recent developmental trends that may be helpful.

Thanks in advance for your kind assistance. More than just seeking your assistance, I also hope to be helpful in offering any pertinent information that would be mutually beneficial to our research directions. I look forward to the opportunity to collaborate with you in the near future.

Sincerely yours,
Danny Wu

【解答】

第一個練習：1. H　2. A　3. G　4. E　5. D　6. B　7. C　8. F

第二個練習：1. I　2. J　3. A　4. G　5. H　6. B　7. C　8. F　9. D　10. E

第三個練習：1. E　2. I　3. C　4. B　5. G　6. A　7. D　8. F　9. J　10. H

第四個練習：1. G　2. B　3 I　4. B　5. C　6. J　7. E　8. D　9. F　10. H

第五個練習：1. F　2. C　3. A　4. B　5. G　6. I　7. D

第六個練習：1. D　2. A　3. F　4. G　5. C　6. B　7. E

第七個練習：1. E　2. H　3.B　4. C　5. A　6. I　7. G　8. F　9. D

【參考文獻】

柯泰德（2006）。《有效撰寫英文電子郵件》。台北：揚智。

Unit Five

Writing Effective Career Statements for Employment

管理師專業英文工作經歷撰寫

簡 介

本單元主要探討管理師如何撰寫專業的英文工作經歷，內容共分為六個部分，分別是：A.表達工作相關興趣；B.興趣相關產業描寫；C.描述所參與方案裡專業興趣的表現；D.描述學歷背景及已獲成就；E.介紹研究及工作經驗；F.描述與求職相關的課外活動。管理師們可以經由參考研究他人寫作範例及相關輔助的練習，逐步強化本身對於英文工作經歷的寫作能力，以成為個人追求專業競爭力的一項利器。

A. Expressing interest in a profession / 表達工作相關興趣

1. Stating how long one has been interested in a particular field or topic / 描述專業興趣所延續的時間，請參考以下範例：

- Although deeply interested in computers since childhood, I gradually became interested in a career in banking out of empathy for the seemingly endless number of potential homeowners who seek quality living accommodations at affordable prices.

- My deep interest in this profession manifested itself in a recent collaboration with urban planners to establish more recreational areas for Taipei urban dwellers.

- A finance course that I enrolled in as an undergraduate student sparked my interest in business management. Becoming aware of how finance significantly impacts our daily lives motivated me to diligently pursue studies in this field.

- Devotion to a career requires compassion, sincerity and determination. Much of my devotion to a career in finance goes back to childhood, when I helped my parents in their small accounting firm.

- My deep concern since childhood over overcrowded urban areas has transformed me into an environmentally conscious individual concerned with equitable living conditions for all Taiwanese residents.

第一個練習：

以下每句爲用來訓練有關表達工作相關興趣：描述專業興趣所延續的時間。請把左邊及右邊的每一片語連結成一個完整句子。解答在本章最後。

1. Integrating many topics such as finance, business administration and sustainable development has	a. and eventually a Master' s degree in the same field from the same institution.
2. Having immersed myself in the strategic management field for quite some time,	b. successful completion of a Master's degree in International Business from Tamkang University.
3. I have aspired to become a financial planner since childhood, explaining why I acquired a bachelor' s degree in the Department of Statistics at National Taiwan Normal University,	c. the Council of Labor Affairs.
4. Having fascinated me since high school, business management encompasses a diverse spectrum of related subjects,	d. to undergraduate school, in which a business management course aroused my curiosity to acquire more knowledge.
5. Commercial development planning has enthralled me since I took part in a business management training course sponsored by	e. enthralled me since undergraduate school.
6. Business-related curricula have fascinated me since I can remember, eventually leading to my	f. since childhood because educators can closely interact with the younger generation.
7. My fascination with quality management can be traced back	g. on in my academic studies fostered my interest in graduate study and a finance-related research career.
8. The teaching profession has strongly interested me	h.with business management-related topics that emphasize sustainable development.
9. While studying in the Department of Business Management as an undergraduate student at National Central University, I became intrigued	i. I am especially interested in researching strategy management-related topics.

10. Exposure to research early	j. including information science, accounting, quality management and statistics.

2.Describing the relevance of one's interest to industry or society / 描述興趣與產業及社會的相關性，請參考以下範例：

· With Taiwan's recent entry in the World Trade Organization, the recent deregulation of Taiwan's market will provide a diverse array of financial services, thus necessitating that domestic companies possess a high level of financial expertise in global markets.

· I remain ever optimistic that quality management approaches can lead to the establishment of senior residential communities that will help care for Taiwan's rapidly aging population.

· Taiwan currently lacks technology professionals who are proficient in the use of such advanced instrumentation, of which, I am quite proficient in using.

· While pursuing this interest, I slowly began to realize how business management encompasses many disciplines, such as finance, statistics and accounting to name a few.

· Although the general public is alarmed over potential radiation harm, carefully controlled amounts of radiation can benefit humans, such as in medical treatment or for industrial use such as construction of residential complexes.

· While cancer ranks as one of the leading causes of mortality in Taiwan, quality management approaches are constantly adopted to monitor the radiation emitted from building construction materials.

3.Stating how one has pursued that interest until now／描述興趣形成過程，請參考以下範例：

- This career aspiration has persisted until now, with my recent completion of a Master's degree in the International Finance from National Taiwan Ocean University.
- Given my interest in this profession, I have collaborated with several information technology firms in developing software programs to incorporate the use of strategic management approaches.
- Having recently completed my master's degree requirements from the Institute of Management Science at National Taiwan University of Science and Technology, I hope to bring my technical expertise and extensive laboratory experience to a renowned consulting firm such as yours.
- To cultivate my interest in this profession, I learned how to perform computer analyses in efficiently handling analysis results, such as compiling statistics to identify unforeseen circumstances in finance projects and to determine the significance of those results.
- The more time spent in studying business administration-related procedures and pertinent industrial-related research has enabled me to more full grasp the application of managerial concepts towards this field.

4.Stating how finding employment related to one's interest would benefit the applicant and/or the company／描述興趣與工作的配合對勞資方都有利，請參考以下範例：

- I am confident of my ability to pursue a research career in this area and, in doing so, significantly contribute to society. Your organization would provide me with such an opportunity.

- If fortunate to secure employment in your organization, I am confident that your company would offer comprehensive training and challenging experiences that would greatly enhance my professional skills. In addition to your solid management training, your company offers an attractive salary and benefits package that I am drawn to.

- I bring to your company a unique background in Chemical Engineering and Business Management, disciplines that will prove to be valuable to any research effort that I belong to.

- I firmly believe that your consulting firm will find my above academic and professional training to be a valuable asset in your highly respected organization.

- My gradual maturation as a proficient researcher during graduate school and subsequent employment will hopefully prove to be a valuable asset in your laboratory, allowing me to contribute significantly to any of your team's research efforts.

第二個練習：

以下每句爲用來訓練有關表達工作相關興趣：描述興趣與工作的配合對勞資方都有利，請把左邊及右邊的每一片語連結成一個完整句子。解答在本章最後。

1.As a member of your organization,	a. but also further enhanced my sensitivity towards various governmental regulations in the finance sector. These are hopefully assets that your research institute is looking for.

2. This diverse education not only strengthened and broadened my solid knowledge base,	b. an attribute which I believe that your organization looks for in its research staff.
3. With my solid academic background and extensive professional experience in Taiwan's banking sector, I feel	c. combination as a valuable asset to any research effort that your staff undertakes.
4. Overall, my diligent study in the management science field has made me more patient and attune to minute details when conducting independent investigations,	d. to further pursue my professional interests.
5. Given my strong academic background and numerous work experiences, I hope to gain	e. offers an excellent environment to further pursue this career path.
6. Given my research emphasis on business management and statistics, I believe that your company will find this unique	f. business transactions models, a system which will equip me with the competence to more significantly contribute to your organization's excellence in marketing.
7. Your consultancy firm would definitely provide me with the opportunity	g. I hope to contribute to your innovative efforts to develop effective management strategies for Taiwan's emerging biotechnology sector.
8. Working in your organization would give me a practical context for theoretical concepts	h. employment in your company owing to its commitment to excellence in quality management. In addition to my technical expertise, I am also interested in enhancing my management proficiency within your organization.
9. I believe that your organization	i. learned in graduate school, giving me a clearer idea of the direction in which my career is taking.

10. I am attracted to your company's advanced financial information system for analyzing	j. equipped to serve as a researcher in management science or accounting-related fields, areas that your company is renowned.

B. Describing the field or industry to which one's profession belongs / 興趣相關產業描寫

1. Introducing a topic relevant to one's profession / 介紹業界所關心的相關主題，請參考以下範例：

· Despite the tremendous number of research personnel in the strategic management field, the role of quality control in the semiconductor industry requires further study, an area of research that I am deeply interested.

· Rapid industrial expansion, environmental pollution, and the increasing quantity of processed foods have all dramatically impacted individual lifestyles, leading to serious illnesses and carcinoma cancer.

· Following its rapid industrial development in recent decades, Taiwan has experienced an unusually high amount of natural calamities and emergency situations. For instance, the mortality rate for traffic accidents in Taiwan is around 62 individuals out of 100,000 involved.

· Regardless of whether for medical care or building construction purposes, measuring radiation dosage levels and providing necessary protection are vital concerns.

· Biotechnology companies and related academic institutions are actively researching ways in which to develop reagents for biochip, gene and protein applications.

2. Describing the importance of the topic within one's profession／強調主題的專業性，請參考以下範例：

- With the increasing number of human diseases, cancer and human genetics research disciplines have subsequently expanded rapidly, raising concern even among urban planners concerned with environmentally friendly management practices.

- For instance, in Taiwan, roughly 34,000 individuals died in 2002 because of malignant diseases. Radiation therapy, surgery, and chemotherapy are three conventional means of treating cancer. 50% to 60% of all individuals diagnosed with cancer will require radiation therapy, some of which are for curative, while others are for palliation.

- Taiwan's current lack of instrumentation and trained personnel explains why no serial study has yet been performed island wide.

- Many Taiwanese researchers involved in developing screen tests lack clinical experience, subsequently preventing the commercialization of product technologies such as those in the semiconductor industry.

- Taiwan's aspiration to join the World Heath Organization necessitates that medical personnel continually upgrade their professional skills — even in broad ranging areas such as radiation monitoring on construction sites and the creation of environmentally friendly spaces for the growing urban sprawl.

3. Complementing the company on its commitment to excellence in this area of expertise／讚美公司對專業的自我期許，請參考以下範例：

- Having frequently held large-scale public discussions on the latest business management trends, your company has equipped itself with state-of-the art equipment and knowledge of pertinent issues for future development.

- As your company continuously develops, technological developments will usher in prosperity, not only for your company but society as well. This explains why I hope to join your corporate family as a researcher.

- As a leader in the quality management field, your company offers an impressive number of wide ranging training courses for your employees so that they remain competent in the field.

- Your company has adopted an integrated approach towards learning by adopting multi-disciplinary approaches towards business management. Such exposure would allow me to interact with professionals from different fields.

- Your engineering firm highly prioritizes offering not only excellent consultancy services for its clients, but also comprehensive on-the-job training for new employees, explaining why I most eager to secure employment at your hospital in a management-related position.

第三個練習：

以下每句爲用來訓練有關興趣相關產業描寫包括：讚美公司對專業的自我期許。請把左邊及右邊的每一片語連結成一個完整句子。解答在本章最後。

1. Your company offers an excellent	a. but also working in a management capacity in the near future.
2. As Taiwan aspires to become a Green Silicon Island', your company's research group has distinguished	b. as evidenced by its creativity in using standard operating procedures to create state-of-the-art product technologies.

3. Moreover, your company's sound reputation can help me to attain my career aspirations of not only enhancing my technical expertise,	c. to develop and commercialize product technologies locally.
4. The quite impressive training courses that your company offers for its employees	d. the opportunity to further pursue my professional interests in the above area.
5. Your company has also distinguished itself in the finance sector,	e. confident of my ability to contribute to your corporate goals.
6. In addition to continually upgrading its operational performance, safety and security,	f. me, given my considerable technical expertise in this field.
7. Your reputed consulting firm would provide me with	g. working environment to pursue this research path.
8. Your company's decision to pursue this line of product technology development is what attracts	h. reflect your excellence in leadership and commitment to staff excellence.
9. As your organization has committed itself to providing quality engineering consultancy services for several years, I am	i. your company concentrates heavily on controlling the institute's property and centralizing data management practices. Those strategies have largely contributed to the solid management team at your company.
10. I am especially impressed with how your company has collaborated with overseas strategic partners	j. itself for its contribution to this field, especially in developing advanced quality management schemes.

4. Stating anticipated contribution to company in this area of expertise if employed there／所期望對公司的專業貢獻，請參考以下範例：

- The opportunity to work at your company would allow me to hopefully contribute to the development of the latest strategic management approaches, hopefully furthering your company to unprecedented levels of growth.

- I believe that your hospital will find my experience in adopting different approaches to enhancing the quality of consultancy services as a valuable asset in strategic planning that is undertaken to enhance your organization's competency in the intensely competitive finance sector.

- Working in your consultancy firm would hopefully allow me not only to pursue some of my above research interests, but also to contribute to the overall welfare of elderly individuals seeking quality living accommodations.

- If successful in securing employment at your company, I bring to your organization a strong scholastic background and practical knowledge, which will hopefully be in line with your company's innovative technological developments.

- If I am successful in gaining employment in your company, both my solid academic training and research on information system development will make me a strong asset in your efforts to upgrade e-business operations, such as in online queries, payments and account transferals.

第四個練習：

以下每句爲用來訓練有關興趣相關產業描寫包括：所期望對公司的專業貢獻。請把左邊及右邊的每一片語連結成一個完整句子。解答在本章最後。

1. I am also confident that my previous work experience will enable me to more fully realize clients'	a. but also enhance my research and knowledge skills.
2. I hope that my entry	b. a valuable opportunity to contribute my background expertise in financial management to your innovative research efforts.
3. Working at your company would not only advance my professional capabilities,	c. efforts to increase managerial efficiency and to optimize the productivity of your company's personnel.
4. If successfully employed at your corporation,	d. me with the necessary skills to provide a diverse array of newly offered services to the community conveniently.
5. Employment at your company would give me	e. needs to maintain their loyalty, an area that your organization is especially concerned with.
6. Following the development of my expertise in business management during graduate school, I can contribute to	f. upon my professional and academic skills in order to become more competent in daily work tasks.
7. Moreover, your company's professional training program for new employees will equip	g. a wealth of technological expertise and experience in the above area.
8. As a member of your construction company, I would continuously build	h. into your organization will ultimately contribute to increased commercial activity and profits.
9. As a member of your corporation, I would bring to your organization	i. I will dedicate my efforts to contributing to the success of the corporate family.
10. Through my previous work and academic experiences, I have acquired specialized skills that	j. make me receptive to novel concepts and diversity in a workplace culture.

C. Describing participation in a project that reflects interest in a profession／描述所參與方案裡專業興趣的表現

1. Introducing the objectives of a project in which one has participated／介紹此方案的目標，請參考以下範例：

· My deep interest in the quality management field led to my active participation in a project aimed at developing a more sound quality assurance system for use in government-sponsored projects.

· My active participation in a project aimed at assessing the quality of a financial consulting firm's services reflects my avid interest in strategic marketing. This undertaking involved collaborating with other marketing planning personnel, quality management employees and educational training staff.

· My recent research emphasis on implementing quality control measures in the banking sector allowed me to grasp the theoretical and practical concepts related to this discipline as well as future trends.

· Although research on quality management practices in long-term residential communities initially began overseas, I was able to extend those results and make them applicable to the needs of Taiwan's growing elderly population.

· Given my commitment to the accounting profession, I actively participated in graduate level research to optimize work productivity in a hi-tech firm, having collaborated with others in related projects.

第五個練習：

以下每句爲用來訓練有關描述所參預方案裡專業興趣的表現包括：介紹此方案的目標。請把左邊及右邊的每一片語連結成一個完整句子。解答在本章最後。

1. I have spent considerable time in project evaluation not only to reduce overhead costs,	a. exposure to strategic management practices in the sustainable development of urban areas.
2. Owing to my deep interest in business management,	b. participated in a project aimed at establishing affordable housing for low-income elderly residents.
3. My recent collaborations have allowed me to gain immense	c. with others in developing a simulation model for non-profit organizations involved in governmental-subsidized projects
4. My recent research effort allowed me to become proficient in not only the use of many radiation detection methods, but	d. in our consulting firm's efforts to accelerating project completion without sacrificing quality.
5. As evidence of my strong interest in this specialized field, I actively	e. but also to avoid the risk of loss valuation.
6. Given my strong desire to thrive within this profession, I collaborated	f. during graduate school attempted to organize an administrative unit that would encourage independently-owned consultancy firms to form a strategic alliance.
7. My sincere commitment to elevating quality control standards in the finance sector is reflected in my active participation	g. so that staff could enhance client services while conserving overhead costs.
8. A proposal for a financial consultancy firm that I wrote	h. also an understanding of how such methods can be applied to evaluated devising quality management guidelines.
9. I recently participated in a cost-benefit analysis project for a consulting firm	i. in which quality control management was of primary emphasis.

10. As evidence of the strong desire to strengthen my research fundamentals, I participated in a management science research project,	j. I am especially interested in adopting strategic management practices in long-term care residential communities.

2. Summarizing the main results of that project / 概述該方案的成果，請參考以下範例：

- Our collaborative efforts led to the discovery of…
- Our results further allowed us to clarify how…
- My recent research enabled me to discern between…
- The results of that project confirmed the effectiveness of our procedure.
- Those project results demonstrated the feasibility of…
- The results of that project confirmed the ability of our system not only to evaluate precisely bottlenecks in industrial management, but also to evaluate immediately the manufacturing system's status with respect to re-scheduling delayed tasks.

3. Highlighting the contribution of that project to the company or the sector to which it belongs / 強調該方案對公司或部門的貢獻，請參考以下範例：

- Results of that research effort clarified/elucidated/shed light on/paved the way for/provided a valuable reference for/facilitated the assessment of/ effectively addressed specific problems relating to/contributed to efforts to…
- Based on the results of that research effort, several consulting firms attempted

to modify their services in order to increase their competence in the highly competitive finance sector.

- Results of that study demonstrated that successfully adopting knowledge management practices among employees enhance the ability of company executives to interact with all staff members.

4. Complimenting the company or organization at which one is seeking employment on its efforts in this area／讚美公司對業界的專業努力，請參考以下範例：

- Your company, a leader in this field of research, could greatly enhance my capabilities in quality management-related topics in the finance sector.

- Given your company's considerable experience in developing quality management strategies for government sponsored projects, I am especially impressed with your ability to evaluate the potential commercialization of various technologies in the Taiwan market.

- Given my professional experiences, your company offers a competitive work environment and highly skilled professionals: these ingredients are essential to continually upgrading knowledge skills and expertise in my chosen career path.

- While offering renowned product brands in the Taiwan market from the United States, your franchise highly prioritizes quality assurance and professionals in a diverse range of management-related fields.

- My profound interest in finance explains my active participation in a project aimed at developing a novel method and promoting public awareness of public sector investment opportunities.

第六個練習：

以下每句爲用來訓練有關描述所參預方案裡專業興趣的表現包括：讚美公司對業界的專業努力。請把左邊及右邊的每一片語連結成一個完整句子。解答在本章最後。

1. Your company's heavy emphasis on ensuring that each new target	a. with other businesses in undertaking innovative public construction projects.
2. Your company can provide me with	b. offers a valuable opportunity for me to advance my knowledge skills.
3. Your company provides a highly collaborative environment	c. upgrade my knowledge expertise and research capabilities through your solid employee training curricula.
4. In this area of development, your company provides a highly	d. reflected in your outstanding human resources and state-of-the-art facilities.
5. Your laboratory, a leader in this field of research,	e. my professional interests.
6. As a leader in the finance sector, your company would provide me with me with many collaborative opportunities to continually	f. can reach its goal successfully ensures a major breakthrough in product technology development.
7. Your commitment to excellence in product innovation is	g. company possesses many highly skilled employees.
8. The opportunity to work at your company would provide me with an excellent environment not only to fully	h. realize my career aspirations, but also allow me to apply theoretical concepts taught in graduate school in the classroom to a practical work setting.
9. Your engineering firm would definitely provide me the opportunity to further pursue	i. a unique opportunity to build upon my previous experiences in the above areas.

10. Besides providing an excellent work environment in the highly competitive banking sector, your	j. competitive work environment and is home to highly skilled professionals.

D. Describing academic background and achievements relevant to employment／描述學歷背景及已獲成就

1. Summarizing one's educational attainment／總括個人的學術成就，請參考以下範例：

- My graduate level research in the Institute of Technology Management at National Chiao Tung University focused on the effectiveness of advanced simulation models in increasing the efficiency of hi-tech administration.

- The nature of my previous work led to my recent completion of a Master' s degree in Business Administration at National Taiwan University, in which I diligently strived to identify what theoretical and academic principles can be applied to my workplace requirements.

- Graduate school allowed me to increase my intellectual capacity by reading the latest management publications, conducting research independently, consulting with financial experts and familiarizing myself with advanced computer languages.

- Having received a bachelor' s degree in Finance and a master' s degree in Business Administration, I am well aware of the theoretical and practical issues involving the assessment of private enterprises in the hi-tech sector.

- My graduate level research in management science focused on increasing on-site productivity, equipping me with sufficient expertise to contribute significantly Taiwan' s hi-tech sector.

2. Describing knowledge, skills and/or leadership qualities gained through academic training／描述學術訓練所獲得的技能及（或）領導特質，請參考以下範例：

- While serving as a research assistant in the Institute of Finance at National Chung Hsing University, I learned how to coordinate different aspects of a research project, whether it be filling out weekly progress reports, managing financial affairs, or organizing regularly held seminars and report contents.

- Despite the rigorous challenges of academic research, my professional experiences reinforce my dedication to laboratory work and ability to develop pertinent research questions and experimental designs, which facilitate data analysis.

- Competency skills such as designing a study, evaluating data and communicating my results orally are essential for civil engineers. Graduate school thus refined my writing and presentation skills while designing a research project.

- With competitiveness among individuals increasing as society advances rapidly, strong analytical skills and the ability to solve problems logically in theory and practice are highly desirable attributes, explaining why I equally prioritize professional and academic experiences.

- Research competency acquired during graduate school definitely enhanced my knowledge of the field, hopefully enabling me to look beyond the role of assisting others in the laboratory to that of leading other research collaborators.

第七個練習：

以下每句爲用來訓練有關描述學歷背景及已獲成就：描述學術訓練所獲得的技能及（或）領導特質。請把左邊及右邊的每一片語連結成一個完整句子。解答在本章最後。

1. Graduate school instilled in me	a. thus motivating me to adopt a more meaningful strategy towards learning, especially how seemingly polar fields can be integrated.
2. Classroom and symposium lectures enabled me to observe how to effectively publish one's research results orally,	b. university instilled in me a theoretical and practical understanding of the latest concepts in business administration.
3. Remaining abreast of current trends in strategic management allowed me to fully engage in discussion with my graduate school advisor	c. including the logical format of the presentation, its deliberate delivery and unique perspectives that are adopted in approaching a research problem.
4. Academic study made me aware that study is more than not just for securing employment,	d. advancing Taiwan's finance sector.
5. My academic advisor encouraged me tremendously, especially when I encountered difficulties during the academic year.	e. but also to conceptualize problems in different ways.
6. Attending several academic conferences and publishing my research findings in internationally renowned journals	f. to understand what research results must be generated in the laboratory to yield a successful outcome.
7. Graduate school equipped me with much knowledge and logical competence so I that I can devote all of my energies to further	g. proficient in using numerous analytical skills, but also nurtured my problem-solving skills.

8. Critical thinking skills developed during graduate studies enabled me not only to explore theoretical concepts and delve into their underlying implications for the accounting profession,	h. the necessity of harmony among research collaborators to achieve a desired outcome.
9. In addition to helping me to nurture skills necessary for the workplace,	i. allowed me to expand the scope of my research activities as well as grasp many helpful concepts on the latest technological trends in civil engineering.
10. Graduate school not only made me	j. This encouragement made me resolute to overcome the rigorous challenges of graduate study, explaining my confidence in persevering despite the intensely competitive nature of the banking sector.

3. Emphasizing a highlight of academic training / 強調一個學術訓練的特定領域，請參考以下範例：

・Although extremely taxing, the project instilled in me a sense of confidence in meeting the rigorous challenges of such an undertaking. Although research of this topic initially began overseas, I was able to extend those results and make them applicable to Taiwan's circumstances.

・I served as a research assistant in the Society of Quality Management, R.O.C., which allowed me not only to see the connection between theoretical knowledge and actual commercial practices, but also to realize how results from academic investigations can influence governmental policies.

・While a course in Statistics focused on how to analyze the effectiveness of the latest software applications, another course in English Writing oriented me on how to organize and write a research paper, as well as increase my academic reading comprehension.

- Given the importance of detail and sequential order when working in a laboratory, I carefully wrote down the details of experimental steps to be followed. Upon completion of the experiment, I compared the data generated from the experiments with those results anticipated beforehand. Despite the occasional bottlenecks in research, I remained calm to avoid experimental failure.

- In addition to providing me with several opportunities to corroborate what I had learned from textbooks in the classroom, several field studies on air pollution at Linyuan Industrial Park in Kaohsiung allowed me to extend my knowledge skills to an entirely different field to achieve innovative solutions.

4. Stressing how academic background will benefit future employment／強調學術背景對未來求職的利處，請參考以下範例：

- Combining academic concepts with my knowledge expertise from previous employment will enable me to rise to the arduous challenges of working in the intensely competitive hi-tech sector for a renowned corporation such as yours.

- The professional knowledge acquired from graduate school will also strengthen my intellectual capacity to give me a clearer picture of my career path, hopefully at your company.

- Feeling somewhat unprepared for the demands of the workplace, graduate school allowed me to further strengthen my knowledge skills in management, thus enabling me to more effectively apply my strong academic background in order to enhance my work competence.

· I am absolutely confident that my academic background and experimental as well as professional experiences in food science and biotechnology will allow me to succeed in your company's many collaborative product development projects.

· Through graduate school research, I contributed to efforts to elevate the quality of affordable housing units for the elderly living in residential communities. I hope to continue this line of work upon employment in your organization.

第八個練習：
以下每句爲用來訓練有關描述學歷背景及已獲成就：強調學術背景對未來求職的利處。請把左邊及右邊的每一片語連結成一個完整句子。解答在本章最後。

1. My outgoing character and aggressiveness towards completing tasks completely	a. during graduate school towards any research effort that I belong to at your company.
2. I will apply the same responsible attitude instilled in me	b. I feel equipped to serve as a researcher in business management or finance fields, areas that your consulting firm is renowned.
3. I would thus like to build upon my solid	c. valuable assets for your company's product development efforts.
4. The opportunity to work at your company would provide me with an excellent environment not only to fully realize my career aspirations,	d.but also allow me to apply theoretical knowledge management concepts taught in graduate school to a practical work setting.

5. I am absolutely confident that my strong academic background will prove invaluable to your company's highly	e. fostered in graduate school when approaching bottlenecks in research during the project implementation phase.
6. I bring to your company the same level of intuitiveness	f. of my professional occupation, but also potential employment opportunities in the rapidly evolving field.
7. Given my solid academic training,	g. to understand any changes that may occur in the business administration field, but also to find answers on how to resolve the problem at hand.
8. Given my strong knowledge skills in analyzing the successful components of a successful management project, I am in a better position not only	h. will definitely prove to be assets in any research effort that I contribute to, hopefully at your company.
9. In addition to the departmental curricula, I fostered practical working skills in order to familiarize myself not only with the demands	i. academic training and relevant work experiences by securing employment in your reputed consulting firm.
10. Hopefully, my strong writing and presentation skills will prove to be	j. demanding product development projects as a researcher who can effectively respond to the latest changes in the rapidly evolving quality management field.

E. Introducing research and professional experiences relevant to employment / 介紹研究及工作經驗

1. Introducing one's position and/or job responsibilities, beginning with the earliest position, and ending with the most recent one / 介紹個人歷年至今所任工作職位及職責，請參考以下範例：

· I served as the chief coordinator of a not-for-profit project, where I was responsible for coordinating personnel and streamlining inter-departmental activities to effectively implement the schedule according to stringent environmentally friendly building code standards.

· When my university department established a continuing educational degree program in 2001, I worked during the daytime and went to class during the evenings. Despite the difficulty of holding down a job and pursuing a bachelor' s degree simultaneously, I acquired much professional knowledge in civil management science.

· My work as an administrative assistant in a consulting firm in Puli of central Taiwan, following my graduation from business vocational school, allowed me to apply theories taught in the classroom to actual working situations.

· After completing our country' s compulsory military service of nearly two years, I began my career working in an information integration company, where I first learned how to maintain hardware systems and construct a network, followed by orientation with the different functions of manufacturing-oriented information systems.

· While diligently striving to familiarize myself with state-of-the-art information technologies, I was eventually promoted to section manager in 1999, a role in which I became adept in accumulating information, evaluating data, negotiating prices and reviewing budgets.

第九個練習：

以下每句爲用來訓練有關介紹研究及工作經驗：介紹個人歷年至今所任工作職位及職責。請把左邊及右邊的每一片語連結成一個完整句子。解答在本章最後。

1. After leaving the semiconductor industry.	a. much theoretical and practical training in practicum internships during the summer band winter breaks.
2. As a graduate student, I acquired	b. the past five years.
3. I worked while receiving my professional license in accounting,	c. explaining why effective communication and a congenial attitude are essential.
4. I have acquired substantial work experience in quality management over	d. I returned to an information integration company, where I was responsible for system analysis as a project manager, acquiring a breadth of experience in dealing with complex management issues.
5. As for my professional experiences, I have acquired much knowledge	e. to commence my career in the finance sector.
6. After receiving my Master' s degree in Business Management from National Tsing Hua University and professional license as an account, I started	f. my horizons as to the dynamics of working in the construction management sector.
7. After completing our country' s compulsory military service of nearly two years, I started working in an engineering consultancy firm	g. enabling me to acquire hands-on and theoretical knowledge simultaneously.

8. The company's comprehensive operations broadened	h. rapidly emerging financial services profession, explaining why I returned to university for a graduate level education.
9. My job demands that I closely interact with others,	i. of quality management-related methods, especially with respect to project management.
10. While finding this work highly stimulating and rewarding, I realized how unprepared I was for the	j. working in an consultancy firm in Taipei, where I received two years of rigorous training on how to effectively implement governmental-sponsored infrastructure projects.

2. Describing acquired knowledge, skills and/or leadership qualities / 描述個人所獲得的知識、技能及（或）領導特質，請參考以下範例：

- Acquiring the above knowledge skills enabled me to more creatively discuss current developments in my research with other collaborators.
- In addition to nurturing my problem-solving and basic knowledge of software engineering, the complexities of interacting with other information specialists profoundly impacted my career direction.
- In addition to refining my ability to coordinate finance projects, I became highly adaptive to change, responsive to sudden fluctuations in user requirements and flexible in acquiring diverse skills demanded in a competitive corporate climate.
- Strategic decision making skills, clear headed thinking, the ability to convey my concerns through effective communication with colleagues and confidence in

myself — all of these attributes allow me to significantly contribute to the task at hand.

- By constantly pursuing my research interests, I try to remain abreast of the latest developments in pertinent literature and strive diligently to grasp their practical implementations.

第十個練習：

以下每句爲用來訓練有關介紹研究及工作經驗：描述個人所獲得的知識，技能及（或）領導特質。請把左邊及右邊的每一片語連結成一個完整句子。解答在本章最後。

1. Although extremely taxing,	a. but also learning how to think independently and deliberately.
2. Related project experiences have greatly strengthened	b. abreast of the latest developments in this area.
3. In addition to refining my ability to coordinate management-related activities,	c. my independent research capabilities and statistical as well as analytical skills.
4. The occasional frustrations of slow progress in research have strengthened my resolve	d. refine my interpersonal skills so that I could effectively lead a development team and achieve on-time delivery of our company's products and services.
5. Laboratory research definitely nurtured my skills that were previously lacking in	e. matured me as an individual.
6. I refined my ability to resolve bottlenecks in research, which required not only absorbing the perspectives of other research collaborators,	f. the project instilled in me a sense of confidence in meeting the rigorous challenges of such an undertaking.
7. More than integrating information technologies in the company through software development, I had to	g. work assignments enabled me to face rigorous challenges of the management sector.

8. By placing a large number of responsibilities on my shoulders, this project greatly	h. to excel in the laboratory, making me more tenacious in spirit.
9. Interacting with other researchers in my field allowed me to remain	i. experimental design, data evaluation and collaboration within a research group.

F. Describing extracurricular activities relevant to employment／描述與求職相關的課外活動

1. Introducing an extracurricular activity that one has participated／曾參加的相關課外活動介紹，請參考以下範例：

・Community service is an integral aspect of scouting activities, with club members looking forward to activities such as working with area youth through physical exercise and extracurricular activities.

・As for my extracurricular activities, I participated in campus wide anti-smoking activities sponsored by Taiwan's Bureau of Health Promotion Department. In addition to offering general orientation on the dangers of smoking, the activities also inform students on tobacco manufacturers' deceptive motives and marketing approaches.

・Most extracurricular activities I participated in were management-oriented out of concern to keep my academic scholarship for studies in business management.

・While certain individuals may view me as being overanxious to assimilate much information, I constantly distinguish between academic/work reading and pleasure reading.

・Getting along with others, especially with classmates in the university's extracurricular activities, was a large part of my daily campus life, regardless of whether I was collaborating on a project or spending leisure time with others.

・My classmates elected me monitor, which is a unique leadership opportunity that increased my self-confidence in leadership skills while striving for harmony among classmates and coordination with the classroom instructor.

・I tried to balance my academic studies with extracurricular activities during university.

・As evidence of my friendly, responsible and communicative nature, I participated in our department's volleyball team during university.

・As for extracurricular activities, individuals need an outlet for expressing their spiritual side and remove themselves from the weariness of daily work.

2. Highlighting the acquired knowledge, skills or leadership qualities that are relevant to employment／強調和求職相關的已有知識技能或領導特質，請參考以下範例：

・Extracurricular activities nurtured my communicative skills by requiring that I closely collaborate with other club members to ensure that everything ran smoothly.

・Organizational responsibilities included drawing up a schedule for management-related activities, encouraging members to actively participate in those activities, leading discussions during meetings and evaluating the success of our activities.

・Scouting activities have profoundly impacted my life, making me more socially conscious, responsible, courageous, independent, aggressive, honest and willing to help others.

- Extracurricular activities and recognition for my contributions greatly added to me character, thus equipping me for the challenges of graduate school and ability to more fully grasp human resource management concepts.
- Language is a vital part of our life's experiences, something that I am devoted to in my leisure time in order to build upon my solid academic background and to acquire a better position at work.

第十一個練習：
以下每句爲用來訓練有關描述與求職相關的課外活動：強調和求職相關的已有知識技能或領導特質。請把左邊及右邊的每一片語連結成一個完整句子。解答在本章最後。

1. Volunteering in the scouting organization greatly impacted my life	a. extracurricular activities enabled me to systematically apply logical reasoning to solve academic problems, such as in operations research, statistics and econometrics.
2. This club not only equipped me with much knowledge expertise on photographing still and moving objects,	b. opinions in order to spur creativity.
3. In addition to equipping me with strong organizational and management skills,	c. accommodate myself within a group to contribute to a team effort.
4. As I enjoy sharing what I have learned with my friends, interacting with others can supplement my	d. opportunities that gave me a broader perspective on the larger global picture.

5. The experience instilled in me the importance of planning in the early stages so as to incorporate all members'	e. to handle personal crises.
6. While training with many classmates from diverse backgrounds, I learned how to easily	f. by making me more conscious of others' needs and aware of my societal responsibilities.
7. Extracurricular activities not only developed my self-reliance skills, but also strengthened my ability	g. knowledge and cultivate interpersonal and communication skills.
8. Extracurricular activities offered diverse educational experiences and opportunities to apply knowledge skills and acquire leadership	h. but also instilled in me an aesthetic quality on how to identify beautiful objects in daily life.

【解答】

第一個練習：1.E 2.I 3.A 4.J 5.C 6.B 7.D 8.F 9.H 10.G

第二個練習：1.G 2.A 3.J 4.B 5.H 6.C 7.D 8.I 9.E 10.F

第三個練習：1.G 2.J 3.A 4.H 5.B 6.I 7.D 8.F 9.E 10.C

第四個練習：1.E 2.H 3.A 4.I 5.B 6.C 7.D 8.F 9.G 10.J

第五個練習：1.E 2.J 3.A 4.H 5.B 6.C 7.D 8.F 9.G 10.I

第六個練習：1.F 2.I 3.A 4.J 5.B 6.C 7.D 8.H 9.E 10.G

第七個練習：1.H 2.C 3.F 4.A 5.J 6.I 7.D 8.E 9.B 10.G

第八個練習：1.H 2.A 3.I 4.D 5.J 6.E 7.B 8.G 9.F 10.C

第九個練習：1.D 2.A 3.G 4.B 5.I 6.J 7.E 8.F 9.C 10.H

第十個練習：1.F 2.C 3.G 4.H 5.I 6.A 7.D 8.E 9.B

第十一個練習：1.F 2.H 3.A 4.G 5.B 6.C 7.E 8.D

【參考文獻】
柯泰德（2003）。《有效撰寫英文工作自傳》。台北：揚智。
柯泰德（2004）。《有效撰寫英文職涯經歷》。台北：揚智。

Appendix

參考範例 i.

Employment Application Statement

My fascination with finance can be traced back to undergraduate school, in which an economics course aroused my curiosity in acquiring more knowledge. With Taiwan's recent entry in the World Trade Organization, the recent deregulation of Taiwan's market will provide a diverse array of financial services, thus necessitating that domestic companies possess a high level of financial expertise in global markets. I am especially interested in those financial aspects that can enhance companies' global competitiveness. Having devoted myself to developing computer information systems for over a decade along with considerable time spent in researching data mining applications in information science for the finance sector, I have developed a particular interest in integrating concepts from these two fields and, then, applying them to the recently emerging financial sector in Taiwan. These two ingredients are definitely crucial to my ability to fully realize my career aspirations, hopefully at your company. Your company is an undisputed leader among financial institutes in Taiwan. Renowned for effectively dealing with unforeseeable emergencies and enhancing customer services, your company has established a vision, which deeply impresses me. Moreover, I am attracted to your company's advanced financial information system for analyzing business transactions models, a system which will equip me with the competence to more significantly contribute to your organization's excellence in marketing.

Rapid advances in information technologies and electronic finance services have become irresistible among consumers in recent years. For instance, local financial institutes have not only developed entirely new approaches and practices, but also face stiff international competition. Thus, relevant research on financial problems and professional training to effectively address such problems has become increasingly prominent. Effectively auditing commercial transactions can prevent the financial system from irregular operations, as well as reduce potential credit risks. Therefore, making information as available as possible is necessary to corporate survival. In this respect, in addition to continually upgrading its operational performance, safety and security, your company concentrates heavily on controlling the institute's property and centralizing traditional banking operation and data management practices.

Those strategies have largely contributed to the solid management team at your company. If I am successful in gaining employment in your company, both my solid academic training and research on information system development will make me a strong asset in your efforts to upgrade e-business operations, such as in online queries, payments and account transferals. Moreover, your company's professional training program for new employees will equip me with the necessary skills to provide a diverse array of newly offered services to the community conveniently.

Owing to a deep interest in how to further strengthen my decision-making capabilities, I initiated a project aimed at developing an efficient product control system capable of monitoring the Work-in-Process (WIP) and dispatching the lot via Automated Material Handling Systems (AMHS). The results of that project confirmed the ability of our system not only to evaluate precisely bottlenecks in production, but also to evaluate immediately the manufacturing system's status with respect to re-scheduling WIP. According to our results, the hypothesis dispatching procedure that we developed increases the performance of AMHS. Additionally, the operation cycle time (OCT) and delivery time (DT) were reduced as well. Your company offers a competitive work environment and is home to highly skilled professionals. The comprehensive and challenging training for individuals involved in upgrading manufacturing technologies would continually upgrade my knowledge skills and expertise in the above area, if given the opportunity to work in your organization.

Having devoted myself to developing information systems in the semiconductor industry for over a decade, I have developed a particular interest in enhancing work productivity via use of the latest information technologies. I have also spent considerable time in researching system integration for manufacturing applications on UNIX-based systems. Critical thinking skills developed my undergraduate and graduate training have enabled me not only to explore beyond the initial appearances of manufacturing-related issues and delve into their underlying implications, but also to conceptualize problems in different ways. Notably, I initiated an intranet project as a section manager at MOSEL Corporation. This responsibility oriented me on how to lead a team and thoroughly understand various software development processes. While participating in several MOSEL group projects, I also learned how to address supply chain-related issues in order to broaden my perspective on potential applications of finance and decision making; these areas are now my main focus of interest. I am confident that my working experience in software development will equip me with the necessary competence to accurately address problems in the workplace. Moreover, my project management experience has enabled me to carefully deal with others and resolve disputes efficiently. My love of challenges will enable me to satisfy constantly fluctuating customer requirements in information integration projects, hopefully at your company.

參考範例 ii.

Employment Application Statement

Commercial development planning has enthralled me since I took part in a business management training course sponsored by the Council of Labor Affairs. I also focused on development planning in medical industry while studying in the Department of Healthcare Management at Yuanpei University of Science and Technology (YUST). These experiences have instilled in me the confidence to undertake a career in development planning. Taiwan currently faces a dilemma in consumer purchasing habits, with an increasing number of individuals switching from pharmacies to merchandisers when purchasing health foods and even medical products. This trend reflects consumer's tendency towards one-stop shopping over merely purchasing specialized products quickly from a convenient location. Given this dilemma, local pharmacies incapable of enhancing their business performance will lose their competitiveness and eventually close. My graduate school research thus focused on how local pharmacies can maintain their competitiveness, in which I identified the primary business factors involved and then developed a management strategy model.

While pursuing a graduate degree, I learned of advanced theories in my field and acquired practical training to enhance my ability to identify and resolve problems efficiently. Moreover, closely studying business practices in the medical sector during undergraduate and graduate school has equipped me with the competence to contribute to the development of management strategies efficient resolution of unforeseeable problems, hopefully at your company. Medical and healthcare expenditures for each Taiwan household skyrocketed 5.4% to 11.5% of entire household income during the period 1991 to 2001. This trend reflects that the life span and health concerns of Taiwan's residents are increasing. Therefore, community pharmacies must strive to make consumer purchasing as convenient as possible. Additionally, this increasing trend medical and healthcare expenditure offers potential profits for Taiwan's community pharmacies. Therefore, most franchisers recruit employees who have experience not only in conducting retail-marketing research, but also planning a series of sustainable business strategies. While pressure to achieve business type diversification of complex drugstores such as Cosmed and Watsons hangs over community pharmacies, chain drugstores have expanded in recent years, as evidenced by Cosmed having 110 units and Watsons having 233 units islandwide as of December 2003. Therefore, community pharmacy chain stores should more heavily emphasize efforts to satisfy consumers in order to raise their market shares as well as enhance competitiveness. As a community pharmacy chain, your franchiser has distinguished itself in overcoming operational difficulties and maintaining good discipline to effectively manage its business units. If I am successful in securing employment at your company, my strong academic and practical knowledge, curricular and otherwise, will enable me to contribute positively to your corporation.

A proposal for a pharmaceutical firm that I wrote during graduate school attempted to organize an administrative unit that would encourage independent pharmacies to form a strategic alliance. In addition to effectively addressing specific problems that each independent pharmacy faces, the proposed unit would also strive to execute joint purchases as well as acquire advanced information technologies that would greatly facilitate the operations of independent pharmacies. Those merits would enhance sales volume, reduce purchasing costs and identify areas of potential growth to gain a competitive edge. Moreover, this proposal examines the feasibility of adopting marketing methods, such as in promotion and advertising. Through the practical experience of proposal and my research focus on the pharmaceutical industry during graduate school, I am confident of my ability to quickly adapt to your company's highly competitive environment. I am even more confident that I can significantly contribute to your franchise. While offering famous medical product brands in the Taiwan market from the United States, your franchise highly prioritizes quality assurance and professionals in a diverse range of medical fields. Specifically, I hope to contribute to your marketing efforts, logistics management and medicine research - areas which are expected to expand in the future.

While pursuing a bachelor's degree in the Department of Health Management at YUST, I received scholarships to support my study and also served as a class officer. Additionally, having my classmates select me as a candidate for the university's exemplary student award affirmed my commitment to studying diligently and fostering my leadership skills. Such experience has significantly prepared me for the rigorous demands of the workplace. During that period, undergraduate coursework instilled in me the fundamentals to conduct research independently. Meanwhile, other courses encouraged classmates to collaborate with each other, thus strengthening my communicative skills greatly. As for knowledge acquired involving the medical sector, several courses such as Health Insurance Systems and Organizational Management in the Health Sector equipped me with the skills to identify and solve problems encountered in Taiwan's national health insurance scheme. I later became adept in adopting marketing and research method such as fuzzy control and the grey theory during my studies in the Graduate Institute of Business Management at YUST. Based on my research findings while applying the gray theory, I submitted an article to the Journal of Health Management on growth trends of pharmacies in Taiwan. My marketing research and highly proficient analytical capabilities will definitely contribute to your company' s commitment to offering quality products and services.

參考範例 iii.

Employment Application Statement

My interest in information technology stems from my first contact in undergraduate computer classes. The Internet especially fascinates me, with its ability not only to provide seemingly unlimited information that can be conveniently accessed anywhere, but also to help me cut down on expenses that I might otherwise spend on textbooks or other literature. Other than the Internet as my information source, I occasionally peruse through the latest technology magazines in the bookstore. Information technologies have definitely transformed living standards globally, as new innovations seem to appear daily. This explains why almost everyone is eager to learn the latest computer programs and acquire as many knowledge skills as possible. Besides my strong interest in information technologies, I have acquired many valuable research experiences through graduate studies in Business Management at Yuanpei University of Science and Technology, These two ingredients are definitely crucial to me fully realizing my career aspirations, skills which I can further refine if employed at your company. The opportunity to work at your company would provide me with an excellent environment not only to fully realize my career aspirations, but also allow me to apply theoretical knowledge management concepts taught in graduate school to a practical work setting.

Taiwan has positioned itself to become the 'Green Silicon Island' following its swift industrial development in recent decades. Besides scientific, technological and societal demands, international marketing and management are essential. However, Taiwan' s information industry severely lacks international marketing and management personnel. Efforts to integrate computers and marketing or management science have recently emerged in Taiwan as a major topic of discussion. Although the Taiwanese information sector has state-of-the-art technologies available, many related products have not been exported abroad. Neither do many Taiwanese products have brand recognition abroad. Thus, local companies aspiring to have an international product appeal should have personnel with international marketing and management backgrounds. As a leader in the information technology sector, your company is renowned for its state-of-the-art products and services, as well as outstanding product research and technical capabilities. As I hope to become a member of your corporate family, the particular expertise developed in graduate school and my strong academic and practical knowledge skills will definitely make me an asset to any collaborative product development effort. Offering more than just my technical expertise, I am especially interested in how your company's marketing and related management departments reach strategic decisions. Employment at your company will undoubtedly expose me to new fields as long as I remain open and do not restrict myself to the range of my previous academic training.

My active participation in a project aimed at assessing the quality of a hospital's medical services reflects my avid interest in strategic marketing. This undertaking involved collaborating with other marketing planning personnel, quality management personnel and educational training personnel. A questionnaire on the level of service quality in the hospital was designed and distributed among patients to understand their attitudes towards the hospital's facilities, instrumentation and software capabilities, which are two important factors as to why they selected the hospital in the first place. In addition to revealing why patients selected the hospital, the results of that project indicated not areas in which patients were dissatisfied with the hospital's facilities, instrumentation and software capabilities. Recommendations were made on what steps hospital administrators should take in remedying these areas of patient dissatisfaction. While customer satisfaction may not necessarily equate customer loyalty, they are directly related. Based on the results of that research effort, the hospital attempted to modify its services in order to increase its competence in the highly competitive medical sector. Related project experiences have greatly strengthened my independent research capabilities and statistical as well as analytical skills. Given my work experiences, your company offers a competitive work environment and highly skilled professionals: these ingredients are essential to my continually upgrading my knowledge skills and expertise in the above area.

I received my undergraduate and graduate training in the Healthcare Management at National Taipei College of Nursing (NTCN). Highly theoretical study and intellectual rigor at NTCN equipped me with strong research fundamentals and knowledge expertise of the medical sector. I later gained entry into the highly prestigious Master's degree program in the Institute of Business Administration at Yuanpei University of Science and Technology. My logical competence and analytical skills scaled to new heights during two years of intensive training. Solid training at this prestigious institute of higher learning equipped me with strong analytical skills and research fundamentals. Graduate courses in Marketing and Statistics often involved term projects that allowed me not only to apply theoretical concepts in a practical context, but also to refine my problem-solving skills. My strong commitment to this field requires acquiring knowledge skills that will not only make me competent, but also allow me to thrive in the workplace. While learning how to adopt different perspectives when approaching a particular problem during undergraduate training, I acquired several statistical and analytical skills during graduate school to solve such problems completely. Doing so involved learning how to analyze problems, identify potential solutions, and implement those solutions according to concepts taught in the classroom. My ability to excel in these areas especially proved effective in devising marketing strategies for research purposes. Despite lacking knowledge expertise of a particular topic at the outset, I strive to quickly absorb new information and adapt to new situations. I believe that your company will find this a highly desired quality.

Unit Six

Writing Effective Work Proposals

有效撰寫英文工作提案

簡 介

國內管理師在撰寫管理英文時常會有的問題，不外乎英文文法錯誤，英文字彙不足或是英文字詞表示不當。然而這些表面的問題都可以自行補救。現在相對重要的一個寫作問題，是許多管理師都是在工作方案完成時，才開始撰寫相關英文提案，他們視撰寫提案為行政工作的一環，只是消極記錄已完成的事項，而不是積極的規劃掌控未來及現在正進行的工作。如果管理師可以在撰寫英文提案時，事先仔細明辨工作計畫提案的背景及目標，不僅可以確保寫作進度、寫作結構的完整性，更可兼顧提案相關讀者的興趣。拙著〈有效撰寫英文工作提案〉即為幫助管理師達成此一寫作目標。

本章主要教導管理師如何建構良好的英文工作提案。內容包括：A.工作提案計畫（第一部分）：背景；B.工作提案計畫（第二部分）：行動；C.問題描述；D.假設描述；E.摘要撰寫：簡介背景、目標及方法，歸納希望的結果及其對特定領域的貢獻；F.綜合上述寫成精確工作提案。

A. 工作提案計畫（第一部分）：背景

1.工作提案建構：你工作提案的主題是什麼？你的讀者可以明瞭工作提案的內容嗎？

範例 i.

As the semiconductor industry rapidly evolves, integrated circuit components continuously develop in line with consumer requirements for thinner and miniaturized electronic products. Wafer chips and system boards have a limited link distance when using a lead frame in conventional semiconductor package technologies, making it impossible for them to conform to high speed communication and high channel density requirements. A collaborative effort between manufacturers of chip packages and carriers has devised a new package technology, i.e., tape automated bonding, a new wafer chip carrier and tape. As a promising alternative to the above limitations, a tape can contain thinner channels with a shorter electrical connection distance between chip and system board to comply with high speed communication and numerous channel requirements.

隨著半導體產業的高度發展，積體電路元件隨著電子產品輕薄短小的需求而不斷地演進，因此封裝製程上面臨了許多的挑戰。原本以導線架爲介面的積體電路構裝方式中，晶片與系統電路板的連接距離有一定的限制，無法符合高速傳遞的電路需求，以及高輸入與輸出腳數的需求。在構裝業者與載板業者的研究與開發下，捲帶式自動接合封裝技術與可撓式軟板就因應而生。由於在可撓式軟板上，可以製作出尺寸更小的電路，因此可以使晶片與系統電路板間形成更短的電性連接距離，可以符合高速傳遞的電路需求，以及高輸入與輸出腳數的需求。

範例 ii.

Time-to-market delivery in semiconductor manufacturing is of priority concern in

advanced R&D technology development. Ensuring that products reach time-to-market delivery goals requires that operational managers and fabrication (fab) personnel fully support R&D experimental lots (R&D Lot). Nevertheless, capacity shortage in a wafer fab fails to comply with output requirements of customers, leading to delays in the R&D lot schedule and the overall project.
對半導體產業而言，新製程的技術開發（Technology Development）是競爭激烈的市場環境下其賴以生存的命脈，爲了能夠及時將產品導入市場，相關管理階層及生產工廠（Fabrication, Fab）皆承諾對研發單位（Research & Development, R&D）及其實驗批貨（Engineering Lot）全力支援。然而，Fab在遇到產能壓力時，往往以今日的利益即量產出貨優先做爲考量，在面臨客户出貨壓力下時，更無力顧及研發實驗而忽略掉研發成果乃公司明日的利益，研發實驗時程或許因此而被延誤，進而造成新產品研發專案延誤。

範例 iii.

Wire bonding of IC packages is critical for both breakdown yield and product reliability.
在IC封裝製程中，銲線管理一直是整體的封裝良率以及IC產品的可靠度的關鍵製程。

2.工作問題：你的工作提案裡有你試著要解決或是想更進一步瞭解的問題嗎？

範例 i.

Process capacity index-Cpk is used to determine the capability performance and equipment stability of the tape manufacturing process. Additionally, process capacity index should, in theory, be positively related to production yield. However, production yield is insufficient even when the tape process has an acceptable Cpk value and stable trend chart in statistical process control. Short, open and etching defects are major defect items identified during analysis of yield loss during final visual inspection.

可撓式軟板製程一向利用製程能力指標來做爲評估製程能力與設備穩定性的依據。理論上來說，製程能力指標的優劣應該和產品良率有直接關聯。然而即便製程能力指標值和管制圖所呈現的製程狀態已經達到穩定，但是產品良率始終無法有效提升。分析造成產品良率不佳的主要不良項目，是與蝕刻有關的短路、斷路與蝕刻不良。

範例 ii.

Although an output-driven fab normally adopts Move and Turn Rates as key performance indicators (KPI), such indicators fail to assess the actual performance of R&D lots in monitoring R&D experiments and ensuring prompt experimental delivery. Fab managers are also interested in the overall R&D cycle time instead of local movement and turn rate indices, subsequently creating a conflict among indices between R&D and fab operations.

Fab以產出爲導向，高度地倚賴晶片移動量(Move)、批貨週轉率(Turn Rate, T/R)做爲主要指標(Key Performance Indicators, KPI)。然而此指標在執行(Run)研發實驗批貨(R&D Lot)方面，卻難以衡量出批貨眞正的績效，不僅不易監控研發實驗進度，亦無法承諾出正確的實驗交期。再者，研發管理階層關心的是總體之研發實驗生產週期時間，而非Move、T/R等局部指標，因而形成指標上的衝突。

範例 iii.

Although intension of first wire bonding has been used to determine the quality of wire bonding, most analyses of wire bonding defects inline and customer complaints with wire bonding defect indicate that the failure location was not located on first wire bonding.

過去銲線製程都是針對第一銲點的拉力強度作爲銲線品質的監控方法，但是實際上的線上銲線不良與客户抱怨的不良現象，大部分都不是發生在第一銲點。

3.問題的量化：你要如何量化問題來讓你的讀者明白之前文獻研究所遇到的量化限制？

範例 i.

For instance, the final visual inspection yield of tape is under 75%. Moreover, major defect rates are -8.5% for short, -4.3% for open and -2.8% for etching processes.

最終外觀檢查的良率低於75%，而主要不良項目的短路不良率爲8.5%；斷路不良率爲4.3%；蝕刻不良率爲2.8%。

範例 ii.

time per mask layer (days). Although managers are also concerned with solutions, the long cycle time of a R&D lot creates vague responsibilities for the wafer fab and R&D, necessitating the development of a feasible cycle time model.

使用傳統方法T/R、Cycle time per mask layer（days）指標來估算研發批貨時間，誤差率經常超過25%。管理者也常思考化解之道，然而，由於研發批貨在生產線上執行的時間經常夾雜了Fab生產因素及研發（RD）實驗因素，當研發批貨延誤或週期時間太長時，責任常常歸屬不清。因此，至今仍找不出合適的有效模式。

範例 iii.

For instance, the defect rate caused by wire bonding was 0.2%, and three customer complaints were related to wire bonding fault in 10Mpcs production output.

目前銲線不良所引起的製程不良0.2%，客户抱怨件數3件，在每月10Mpcs。

4. 問題的中心：如果問題沒被解決或是充分瞭解，這對工作提案的讀者會有多大的負面衝擊？參考範例：

範例 i.

The inability to enhance the production yield of the tape process makes it impossible to achieve a satisfactory production yield owing to the lack of a good Cpk, ultimately incurring high production costs and low product quality.

如果無法找出提升Tape生產良率的解決方案，對於研究單位則無法得知為何在Tape製程上，良好的製程能力指標為何無法擁有合理的高產品良率。此外對於業界而言，低落的產品良率無法在成本上與品質上形成足夠的競爭力，製造公司將無法生存。

範例 ii.

As for the total cycle time of a R&D lot, TCTRD, consists of fab-run RD lot time (TCTFab) and R&D development handling time (HT_RD). Restated, TCTRD= TCTFab + HT_RD. Vague responsibilities and inadequate indices may delay the lot schedule and impede time-to-market delivery of advanced technology products.

由於研發批貨總生產週期時間（TCTRD）係由Fab執行研發批貨所花費的時間（TCTFab），加上研發人員相關的處理時間（R&D Development Handling Time, HT_RD）所組成。責任歸屬不清及缺乏合適的衡量指標情況下，導致批貨交期延誤，更進而造成研發新製程技術開發無法及時上市。

範例 iii.

The inability to identify the root cause of wire bonding defects and eliminate them makes it impossible to identify the optimal process parameters, exacerbating the high wire bonding defect rate and customer complaints.

如果無法找出銲線不良的原因並消除，對於研究單位則無法得知銲線製程未最佳化之處。對於業界而言，銲線不良與客户抱怨會造成持續的內外部品質成本以及商譽的嚴重損失。

5.工作提案計畫（第一部分）：背景

範例 i.

Setting of work proposal工作提案建構 The Taguchi method is extensively adopted to enhance manufacturing procedures and product quality. As an efficient robust design approach, the Taguchi method significantly enhances process performance and reduces process development time and manufacturing costs. Additionally, the Taguchi method uses parameter design to determine a robust factor-level combination that can intervene in noise factors. Noise factors produce product/process variances, making such factors extremely difficult or costly to control. Moreover, experiments allocating Taguchi's orthogonal array can reduce the quantity of experiments to lower experimental costs. The Taguchi method has subsequently been adopted in numerous industrial-related fields. Work problem工作問題 In most design instances, increasing variation in customer requirements accompanies complex product design. Although adopted in various industries to continuously enhance product design in response to customer requirements, the conventionally adopted Taguchi method can only optimize single quality characteristic design. As product/process design tends to be rather complex to comply with constantly changing customer requirements and production technologies, quality improvement of the conventional Taguchi method gradually declines. Quantitative specification of problem 問題的量化 The Taguchi method applies a signal-to-noise (SN) ratio as a quality performance measure. According to the SN ratio calculated from experimental observations, an optimal factor-level combination can be obtained by selecting the factor-level with a maximum SN ratio. However, utilizing the SN ratio as a quality performance measure in the optimization procedure may yield an erroneous analysis. Furthermore, the elevated degree of SN ratio for the optimal factor-level combination deteriorates when a complex relationship exists between control factors and quality characteristics. Importance of problem問題的中心 Manufacturers are especially concerned with optimizing a system's quality to enhance product competitiveness in the market place. Most recent manufacturing systems have a complex design with multiple quality characteristics based on various customer requirements. Accordingly, multiple quality characteristics must be evaluated simultaneously to determine an optimal factor-level combination for a system. The inability to efficiently optimize a complex system with multiple quality characteristics limits applications of the conventional Taguchi method.

範例 ii.

Setting of work proposal 工作提案建構 Whereas product quality in manufacturing is often evaluated using process capability analysis, the process capability index (PCI) assesses product quality via quantification measurements. Consequently, process yield is closely related to PCI. In addition to providing valuable process or product-related information for quality control engineers, process yield reflects the degree of product quality in an enterprise. Work problem工作問題 As is well known, Cp, Cpk and Cpm are commonly used PCIs in manufacturing. While the sample statistic is utilized to approximate the population parameters, and are not the unbiased estimators of Cp, Cpk and Cpm. Additionally, the probability density functions (PDFs) of these indices are difficult to obtain, thus complicating the computing task of the confidence interval for these indices. Quantification of problem問題的量化 Otherwise, the most commonly used index Cpk has an inequality relationship, not a one-by-one relationship, with process yield. Importance of problem問題的中心 The inability to achieve an equality relationship makes it impossible to construct the confidence interval of process yield through Cpk and other indices.

B. 工作提案計畫（第二部分）：行動

1.工作目標：工作提案的目標？

範例 i.

Based on the above, we should identify the root causes incurring low production yield in tape manufacturing by analyzing process parameters and possible causes of major visual defects.

基於以上所述，我們應該檢視製程參數以及主要的外觀不良的成因，期望找出產品良率低落的原因，在現有設備製程條件下加以改善。

範例 ii.

Based on the above, in addition to constructing an effective performance index, we should develop an optimal R&D lot cycle time reduction model, capable of executing timely, effective and clearly defined measures to achieve project cycle time, regardless of whether in R&D or in a wafer fab.

基於以上所述，我們應該發展出有效的績效指標及適合研發批貨生產週期的運作模式，模型需能及時執行、有效，並且對Fab及研發單位所做的生產週期努力均能清楚定義。

範例 iii.

Based on the above, we should develop a measurement method, capable of determining the process capability, thus reducing the defect rate and customer complaints by optimizing the process parameters.

基於以上所述，我們應該檢視量測方法，評估銲線製程的眞實製程能力，再進一步做到製程參數最佳化，以降低其不良率及客户抱怨的發生。

2.達成目標的方法：你的計畫中達成目標的步驟？請參考以下範例：

範例 i.

To do so, the leading defect items can be determined from production yield data. Possible causes of major visual defects can then be simulated based on the cause and effect diagram method. Next, parameter combination from crossing processes can be optimized by using the Taguchi method to reduce the defect rate of the tape process.

從現有良率的損失資料統計前三大不良項目，再從主要外觀不良項目以特性要因法推導可能的製程原因，再藉由田口方法從不同的製程參數組合中找出能獲得最高產品良率的跨製程參數組合。

範例 ii.

To do so, X-factors can be modified to determine the R&D cycle time and derive the F-factor. Based on the two factors, a model can then be constructed to shorten technology development time and continuously improve quality control by using the SPC chart as a monitoring mechanism. Next, the X-factor (XFab), excluding R&D handling time (HT_RD), can be used to determine the on-time delivery performance of a wafer fab for an advanced R&D technology. Additionally, the F-factor can be used to monitor the maturation process of advanced R&D technology manufacturing and reduce lot inventory costs.

若要這麼做，以修正的X-factor來計算研發批貨時間，並發展出績效衡量指標，依此建構模型以降低研發生產週期，此外，管制衡量指標以做持續性的改善。模型中以X-factor指標衡量Fab在執行研發批貨上的準時達交程度，同時以創新的F-factor指標評估研發單位是否及時解決製程上的問題，此F-factor指標並可做爲研發技術移轉時製造成熟度及庫存成本。

範例 iii.

To do so, a measurement method for wire bonding tension can be reviewed based on replies from the wire bonding defect inline and analysis of customer complaints. Process capability can then be estimated accurately. Additionally, process parameters can be optimized by using the Taguchi method to enhance quality of wire bonding.

以現有生產線上的銲線不良與客户抱怨的不良現象與位置，檢討銲線強度量測方法，從量測方法的改變以精確得知實際的製程能力，再藉田口方法尋求最佳的製程參數組合。

3.希望的結果：你希望達成的結果？

範例 i.

As anticipated, the proposed Taguchi method can reduce the occurrence of the leading three defects by 10% and increase the production yield above 80%.

希望本研究能降低tape製程前三大不良率的10%，並提升整體產品良率至80%。

範例 ii.

As anticipated, the proposed performance indices can be used to accurately forecast the R&D lot schedule, shorten lot cycle time and ensure time-to-market delivery of advanced technology products. KPI design provides global optimization benefits that link R&D and the wafer fab.

如所預期的，績效衡量指標除了正確預測研發批貨生產週期之外，所提出的模型亦能縮短研發生產週期，進而確保新製程技術能夠及時上市。衡量指標的設計，期望化解研發與管理及製造單位共用使同一Fab的衝突，以達公司整體績效最佳化。

範例 iii.

As anticipated, the proposed measurement method can reduce the wire bonding defect rate from 0.2% to 0.1%.

希望本研究能降低銲線製程不良率至0.1%。

4.領域的貢獻：你的提案對相關工作領域的貢獻？

範例 i.

Results of this study can not only provide further insight into how process performance index and production yield in the tape process are related, but also contribute to efforts to increase production yield and enhance industrial competitiveness.

對於研究單位可以得知爲何在Tape製程上，良好的製程能力指標與合理的產品良率的關係。對於產業界而言，可以找出提升Tape生產良率的解決方案，提升產品良率與品質，以形成足夠的競爭力。

範例 ii.

The proposed performance index can ultimately shorten R&D cycle time and enhance the on-time delivery of advanced technology products by fab-run R&D Lots. Given R&D technology trends to enhance product quality and manufacturing maturity through KPI design, enterprises can exploit these factors to upgrade operational performance in semiconductor manufacturing.

本研究的貢獻在於找出研發實驗批貨生產週期的方法論、設計出績效衡量指標，依此來建構研發生產週期降低模式。此模型可有效縮短先進製程開發時間，進而產品及時上市，提升半導體產業公司的營運績效及競爭優勢。

範例 iii.

In addition to contributing to efforts to develop a feasible measurement method of wire bonding tension, the proposed method can simulate the root cause of wire bonding defects, ultimately providing a solution for rising wire bonding tension in the IC packaging industry in order to enhance assembly yield and wafer quality. 對於研究人員可以研究出正確的銲線強度量測方法以及銲線不良的真正原因。對於產業界而言，可以找出提升銲線強度的解決方案，並提升產品良率與品質。

5.工作提案計畫（第二部分）：行動

範例 i.

Work objective工作目標 Based on the above, we should develop a product/process optimizing procedure based on ABC method, capable of efficiently optimizing a system regardless of the system complexity. Methodology to achieve objective 達成目標的方法 To do so, experimental observations of quality characteristics for a system can be formed based on statistical approaches. Given the ability to simultaneously evaluate multiple inputs and outputs for a system, quality performance of the system can then be assessed using the ABC method. Next, a solving scheme can be used to obtain the optimal factor-level combination of the complete system. Additionally, a verification process can be added to the optimizing procedure in order to verify the quality improvement of the optimal factor-level combination. Anticipated results希望的結果 As anticipated, capable of alleviating the limitations of the conventionally adopted Taguchi method, the proposed procedure using ABC method does not require specific assumptions regarding a system, allowing it to efficiently determine an optimal factor-level combination for a system. Additionally, statistics can enhance the solving efficiency for the optimizing procedure. The proposed procedure can also significantly enhance quality performance regardless of design complexity. Restated, the proposed procedure can obtain a robust design for a system. Contribution to field領域的貢獻 The proposed product/process optimizing procedure using ABC method can contribute to efforts to continuously improve the conventional Taguchi method and optimization schemes for multiple quality characteristics, subsequently helping manufacturers to optimize recent complex design systems to satisfy diversified customer and technology requirements. Furthermore, the proposed procedure can obtain a factor-level combination more robust than the conventional Taguchi method, enabling manufacturers to earn a profit from quality improvement.

範例 ii.

Work objective 工作目標Based on the above, we should develop a standardized procedure to determine the process yield through PCI in order to generate the confidence interval. Methodology達成目標的方法 To do so, the confidence interval of process yield can be constructed based on the index Spk, which is defined by Boyles (1994) as having a one-by-one relationship with process yield. The confidence interval of process yield can then be approximated according to Spk by adopting a repeated sampling method called Bootstrap Simulation. Next, reliability of the simulation results can be confirmed by performing verification analysis. Additionally, the effect of various combinations of simulation parameters can be determined using sensitivity analysis. Anticipated results 希望的結果As anticipated, the proposed standardized procedure involves use of a Visual Basic-based application program for engineers without a statistical background. Contribution to field 領域的貢獻Importantly, in addition to providing another means of obtaining the confidence interval of process yield based on bootstrap simulation, the proposed method can be easily implemented in enterprises to determine product more precisely and facilitate decision making.

C. 問題描述

1. 工作提案建構：你工作提案的主題是什麼？你的讀者可以明瞭工作提案的內容嗎？

範例 i.

Roughly 90% of all Taiwanese enterprises are small or medium-sized. As lending and investing are common business practices of financial organizations, banks must determine whether borrowing enterprises are worth investing in. In developing a credit scoring model for enterprises, banks must accumulate pertinent data such as financial sheets and indexes. Given such disorganized data, banks rely on financial experts to determine whether the import attributes are input attributes.

範例 ii.

IC consumers increasingly demand the semiconductor backend turnkey service (SBTS) offered by wafer-fabrication manufacturers. Those manufacturers integrate their efforts with semiconductor backend original equality manufacturers (OEM) to execute entire backend operations, including probing testing, assembly and final testing. In addition to the ability to remove numerous orders releasing and tracking from clients, SBTS assigns the responsibility of outsourcing allocation and order coordination to the wafer fab manufacturer. A production planning approach for SBTS should thus be devised to comprehensively address the problem among multi-products, multi-stages and multi-sites. In addition to production planning, transportation-related issues should also be considered to enhance the reliability of such an approach. However, solutions for the complex model are inefficient with an expanding scale of the problem, thus necessitating the development of an efficient method for the SBTS model.

範例 iii.

Inventory refers to the stock of an organizational-related item or resource. All firms maintaining an inventory supply can satisfy consumer demand with respect to product variation. However, a large inventory may incur long product cycle times. Additionally, the proportion of inventory and setup costs of a production line to the total operational fund in a company is extremely high, highlighting the importance of developing a more economic inventory order model to satisfy production requirements.

2. 工作問題：你的工作提案裡有你試著要解決或是想更進一步瞭解的問題嗎？

範例 i.

However, the ability of financial experts to determine the correct inputs makes it impossible for the classifier to achieve correct outcomes. In credit scoring research, assuming that all attributes are input data is not always a proper strategy since improper input classifiers often result in imprecise classification models. This is owing to that improper inputs interfere with classifiers in developing the correct mathematical relation between data. Until the attribute selection module is considered, classifiers can train stable and dependable models by using important attributes.

範例 ii.

Although integer linear programming (ILP) is a conventional means of modeling and resolving problems, the iteratively approximation method appears to be inefficient while the problem tends to be a NP-hard or NP-complete one. While Leachman (1992) simplified the ILP model of capacity allocation problem by Lagrange’s relaxation, those results merely indicated the boundaries of the optimized solution, not the real ones. Some studies adopted artificial intelligent approaches, e.g. genetic algorithm, simulation anneal algorithm and neural network, to acquire nearly optimized solutions efficiently. Nevertheless, such approaches must be implemented under specific model types. The heuristic algorithms developed for this unique problems lacks extensive applications, making it extremely difficult to control the performance accurately.

範例 iii.

The economic production quantity (EPQ) model derives an optimal production lot size that minimizes overall inventory costs for a single item. However, a situation when multiple items are scheduled on a single facility does not guarantee a feasible EPQ solution for each item in order to prevent stock depletion during the production cycle. The rotation cycle policy assumes that exactly one setup is available for each product during each cycle.. Additionally, all products are manufactured in the same sequence during each production cycle. Despite its feasibility, the rotation cycle approach is not optimal in terms of minimizing overall production costs.

3. 問題的量化：你要如何量化問題來讓你的讀者明白之前文獻研究所遇到的量化限制？

範例 i.

For instance, a model incapable of selecting important attributes leads to unreliable classification models.

・For instance, a scheduling problem involves identical parallel machines in which ten products, ten photocopy machines and 100 lots in the TFT array process require at least ten hours for solving the scheduling problem via integer linear programming. According to previous estimates, resolving production planning-related problems among multi-products, multi-stages and multi-sites requires more than one week.

範例 ii.

For instance, for product m1 with a production time twice that of m2, the rotation cycle obtains a feasible solution. Both products are manufactured once in a single cycle, with the total cost significantly higher than the sum of two single optimal solutions.

4. 問題的中心：如果問題沒被解決或是充分瞭解，這對工作提案的讀者會有多大的負面衝擊？

範例 i.

Moreover, data mining research is limited in scope with respect to credit scoring methods.

範例 ii.

The inability to simultaneously consider multi-products, multi-stages and multi-sites implies that the optimized solution simply refers to the local optima, not the global one. Furthermore, the inability to resolve the problem efficiently makes it impossible to implement a time-consuming production planning mechanism.

範例 iii.

A rotation cycle policy is developed to modify EPQ when multiple items are scheduled on a single facility, as well as to vary the production time of all products to the same frequency. Doing so can lead to the establishment of a simple and accessible formula. However, when products vary too widely in terms of manufacturing time, adopting the rotation cycle policy creates a feasible solution, but not an optimal one, leading to a markedly higher cost as well.

5. 計畫需求：根據以上問題，最迫切的計畫需求是什麼？

範例 i.

Therefore, a feature selection-based operating model must be developed, capable of obtaining stable and dependable credit scoring models by selecting more powerful attributes as well as considering and storing more information than conventional models can.

範例 ii.

Therefore, a production planning model for SBTS must be developed to simultaneously consider the attributes of multi-products, multi-stages and multi-sites. A TPP network transformation approach must also be developed to apply TPP heuristic algorithms for solving SBTS efficiently.

範例 iii.

Therefore, a process optimization procedure must be developed, capable of modifying the production cycle time based on the rotation cycle method.

6.問題描述，請參考以下範例：

Setting of work proposal 工作提案建構The Taguchi method is extensively adopted to enhance manufacturing procedures and product quality. As an efficient robust design approach, the Taguchi method significantly enhances process performance and reduces process development time and manufacturing costs. Additionally, the Taguchi method uses parameter design to determine a robust factor-level combination that can intervene in noise factors. Noise factors produce product/process variances, making such factors extremely difficult or costly to control. Moreover, experiments allocating Taguchi's orthogonal array can reduce the quantity of experiments to lower experimental costs. The Taguchi method has subsequently been adopted in numerous industrial-related fields. Work problem 工作問題In most design instances, increasing variation in customer requirements accompanies complex product design. Although adopted in various industries to continuously enhance product design in response to customer requirements, the conventionally adopted Taguchi method only can optimize single quality characteristic design. As product/process design tends to be rather complex to comply with constantly changing customer requirements and production technology, the quality improvement of the conventional Taguchi method gradually declines. Quantitative specification of problem 問題的量化The Taguchi method applies a signal-to-noise (SN) ratio as a quality performance measure. According to the SN ratio calculated from experimental observations, an optimal factor-level combination can be obtained by selecting the factor-level with a maximum SN ratio. However, utilizing the SN ratio as a quality performance measure in the optimization procedure may yield an erroneous analysis. Furthermore, the elevated degree of SN ratio for the optimal factor-level combination deteriorates when a complex relationship exists between control factors and quality characteristics. Importance of problem問題的中心 Manufacturers are especially concerned with optimizing a system's quality to enhance product competitiveness in the market place. Most recent manufacturing systems have a complex design with multiple quality characteristics based on various customer requirements. Accordingly, multiple quality characteristics must be evaluated simultaneously to determine an optimal factor-level combination for a system. The inability to efficiently optimize a complex system with multiple quality characteristics limits applications of the conventional Taguchi method. Project need 計畫需求A novel robust design method must therefore be developed to overcome the limitations of the Taguchi method.

D.假設描述

1.工作目標：工作提案的目標？

範例 i.

A feature selection-based operating model can be developed, capable of obtaining stable and dependable credit scoring models by selecting more powerful attributes as well as considering and storing more information than conventional models can.

範例 ii.

A production planning model for SBTS can be developed to simultaneously consider the attributes of multi-products, multi-stages and multi-sites. A TPP network transformation approach can also be developed to apply TPP heuristic algorithms for solving SBTS efficiently.

範例 iii.

A process optimization procedure can be developed, capable of modifying the production cycle time based on the rotation cycle method. During each cycle, products may have more than one setup. A solution procedure can determine the appropriate cycle length, number of setups for each item, scheduling constraints and total inventory cost function. Via this procedure, each item cycle can be close to the prime solution derived from the EPQ model.

範例 iv.

To do so, efficient attributes can be selected using the feature selection method. More information can then be stored in the fuzzy data set. Next, classification can be defined with a two-stage clustering method that combines a self-organizing map (SOM) and K-Means. Additionally, the number of cluster and initial points of K-Means can be determined with SOM. Moreover, clusters can be labeled by K-Means. Also, an inconsistent cluster and isolated instance can be detected with two-stage clustering. Furthermore, the classification model can be obtained from ANN training.

範例 v.

To do so, an ILP model can be devised for SBTS. The Lingo package software can then be equipped to resolve the ILP problem. Next, the SBTS problem can be transformed to a TPP network problem while the ILP becomes inefficient. Additionally, the TPP network problem can be resolved rapidly by implementing the corresponding heuristic algorithms. Moreover, the solutions from the TPP heuristic algorithms can be projected to the original SBTS problem, with feasible results of SBTS obtained as well. Furthermore, the TPP heuristic algorithms can be compared to identify the most qualified one to implement a significant number of cases.

範例 vi.

To do so, the model framework can be devised in contrast with the rotation cycle policy. The policy can be adjusted to identify how each item is related with respect to optimal cycle time. Feasibility of the proposed model can then be verified, with desired changes made accordingly. Next, a recycle time formula, conditions of model constraints and total cost formula can be established. Examining the rule by iteration can allow us to analyze the rationality and validity of these models. Additionally, an actual example can be applied to demonstrate its feasibility, along with a comparison made to the solution of the primal rotation cycle. Moreover, sensitivity analysis can be performed to demonstrate how alternation cycle time impacts total cost.

2.希望的結果：希望的結果你希望達成的結果？請參考以下範例：

範例 i.

As anticipated, the proposed feature selection-based operating model is more efficient and stable than the conventional approach of selecting attributes with experts. Additionally, the proposed hybrid method can construct a classifier applicable to credit scoring of small- and medium-sized enterprises more economically.

範例 ii.

As anticipated, the proposed SBTS-based production planning model can accurately reflect the total cost among multi-products, multi-stages and multi-sites by simultaneously considering production planning and transportation factors. Based on those results, potential OEMs can be identified as strategic alliances to facilitate long-term cooperation. Moreover, adopting the proposed TPP network approach can reduce at least 50% of the resolution time, thus enabling wafer fab manufacturers attempting to adopt SBTS to quickly respond to client requests regarding receiving orders.

範例 iii.

As anticipated, the proposed algorithm can resolve the EPQ and rotation cycle problem of multiple items for a production facility. Numerical analysis of a case study demonstrates that the proposed procedure can reduce overall cost more than rotation cycle policy can. In doing so, the setup cost and setup time can be integrated into the proposed algorithm. Additionally, production line facilities can be upgraded by centralizing the idle times simultaneously, thus ensuring their efficiency, e.g., routine maintenance and flexibility in production scheduling.

3.領域的貢獻：你的提案對相關工作領域的貢獻？

範例 i.

Importantly, the proposed method can prevent trivial attributes from interfering with important ones, as well as moderate them.

範例 ii.

The proposed TPP network transformation approach can resolve large scale LIP problems efficiently. In contrast to other algorithms, the proposed approach tends to transform the original problem into a correlated TPP network one. TPP heuristic algorithms can thus be utilized to ensure efficiency. Moreover, the production planning model for semiconductor backend turnkey services facilitates wafer fab manufacturers in acquiring complex information to achieve economic production.

範例 iii.

The proposed method can resolve problems involving multiple items for a facility. Consequently, production management can construct a feasible solution under multiple items on a facility. Additionally, a larger ratio for optimal production time for one item to another implies more benefits acquired from the proposed algorithm. Determining production quintiles is often referred to as a trade-off problem, in which the formulae attempt to minimize cost, reduce inventory cost, enhance product quality and increase corporate profits.

4.假設描述，請參考以下範例：

Objective of work proposal 工作題案目標 A product/process optimizing procedure based on ABC method can be developed, capable of efficiently optimizing a system regardless of the system complexity. Methodology to achieve objective 達成目標的方法？To do so, experimental observations of quality characteristics for a system can be formed based on statistical approaches. Given the ability to simultaneously evaluate multiple inputs and outputs for a system, quality performance of the system can then be assessed using the ABC method. Next, a solving scheme can be used to obtain the optimal factor-level combination of the complete system. Additionally, a verification process can be added to the optimizing procedure to verify the quality improvement of the optimal factor-level combination. Anticipated results 希望的結果 As anticipated, capable of alleviating the limitations of the conventionally adopted Taguchi method, the proposed procedure using ABC method does not require specific assumptions regarding a system, allowing it to efficiently determine an optimal factor-level combination for a system. Additionally, statistics can enhance the solving efficiency for the optimizing procedure. The proposed procedure can also significantly enhance quality performance regardless of design complexity. Restated, the proposed procedure can obtain a robust design for a system. Contribution to field領域的貢獻The proposed product/process optimizing procedure using ABC method can contribute to efforts to continuously improve the conventional Taguchi method and optimization schemes for multiple quality characteristics, subsequently helping manufacturers to optimize recent complex design systems to satisfy diversified customer and technology requirements. Furthermore, the proposed procedure can obtain a factor-level combination more robust than the conventional Taguchi method, enabling manufacturers to earn a profit from quality improvement.

E. 摘要撰寫：簡介背景、目標及方法，歸納希望的結果及其對特定領域的貢獻

1. 工作提案建構：你工作提案的主題是什麼？你的讀者可以明瞭工作提案的內容嗎？

範例 i.

With the unique environment of Taiwan's National Health Insurance scheme and trend for hospital clinics to outsource their medical information needs to information technology vendors, appraising potential vendors to outsource those needs is extremely difficult, especially given the current ability of medical information systems operating under the National Health Insurance scheme to select the most reliable information technology supplier.

範例 ii.

Outsourcing essential medical services is a growing trend among healthcare institutions. Following budgetary reform of the National Health Insurance Bureau in Taiwan, besides developing many private medical services, hospitals have increasingly outsourced the services of medical examination centers as a vital source of income. Among the many ways of outsourcing advanced medical examination centers include variations on aspects such as hospital management strategies, medical instrument and material suppliers.

範例 iii.

Logistics centers play a pivotal role in ensuring that the importing and exporting activities of international firms are implemented efficiently, ensuring that commercial trade can satisfy consumer demand and offer quality services.

2. 工作問題：你的工作提案裡有你試著要解決或是想更進一步瞭解的問題嗎？

範例 i.

Consequently, the inability of hospitals to adopt an explicit and objective standard in order to select information technology suppliers may lead to an inappropriate selection, subsequently increasing development and maintenance-related costs, a waste in human resources and time, inefficiency in the processing of medical expenditures, as well as the inability to promote the quality of medical treatment.

範例 ii.

Additionally, the organizational framework and attitudes of medical professionals cause conflicts due to the competitiveness and differences in developmental stages, outsourcing models and service domains among hospitals. Moreover, outsourcing the services of a medical examination center is complex and different from other services owing to the involvement of professionals and the continuous requirement to maintain medical service quality. Constrained by governmental procurement and medical guidelines, hospitals find it extremely difficult to select an appropriate outsourcing candidate based on their needs, necessitating the development of a model to select the most qualified outsourcing supplier.

範例 iii.

However, conventional management practices of logistics centers have difficulty in complying with the demands of local enterprises, negatively impacting the quality of services delivered owing to the inability to effectively respond to corporate changes such as the demand for increasingly smaller products and materials, a shorter product life cycle and limited time in ordering and purchasing goods.

3.工作目標：工作提案的目標？

範例 i.

Therefore, this work presents an AHP-based decision-making method for hospital clinics to objectively assess the quality of an information technology supplier when outsourcing their medical information needs, as an alternative to previous decision-making approaches based on subjective evaluations.

範例 ii.

Therefore, this study develops an evaluation model to identify appropriate outsourcing candidates.

範例 iii.

Therefore, this work presents a flexible and accurate fuzzy theory-based AHP method that applies fuzzy theory to the business ratings of logistics centers, providing corporate managers with an effective criterion to assess the management practices of logistics centers and the quality of services delivered.

4.達成目標的方法：你的研究中，達成目標的步驟？

範例 i.

An appraisal criterion for selecting an appropriate information technology vendor is made using the Delphi method to gain mutual recognition among experts in the field. Criteria for outsourcing vendors and sub-criteria of medical information system for clinics are then evaluated using the group decision process of the Delphi method. Next, a questionnaire is designed for constructing the decision attributes in order to evaluate MIS vendors. Additionally, via use of the Delphi method, preferences of the group members are elicited and a group consensus is achieved. Moreover, results of interviews with experts are integrated with the AHP method to establish an objective evaluation criterion for MIS vendors. Furthermore, an AHP-based survey is designed using pairwise comparison to respond to MIS outsourcing demand scores for each item. Also, each criterion and rank can pass a consistency test for decision-making of the AHP model. Finally, participation and involvement of the group experts combined with AHP are incorporated to calculate the weights and optimum alternatives.

範例 ii.

Location-related factors are identified through an exhaustive literature review and consultation with field experts. Exactly how these individual factors are related to each other is then identified through use of AHP. Next, a questionnaire is submitted to hospital procurement professionals to examine the correct factors. Additionally, all factors involved in selecting the target population and factories and stores are analyzed using AHP, with the optimal factories and stores chosen based on those factors.

範例 iii.

A questionnaire is designed based on interviews with managers of logistics centers. Data obtained from those interviews regarding service quality are then analyzed using the AHP method and fuzzy theory. Next, service quality of the logistics centers is determined using an AHP-based service quality and assessed criteria. Additionally, quality of service is assessed based on the corporate expectations as the enterprise-like corporation. Moreover, customer service and other special services are incorporated into efforts to achieve customer satisfaction.

5.希望的結果：你希望達成的結果？

範例 i.

The proposed AHP-based decision-making method can provide hospital administrators with a decision-making and evaluation criteria that actively encourage the medical sector outsource to information technology contractors.

範例 ii.

Analysis results can be used as a model for selecting the most qualified outsourcing company. Adopting the AHP-based assessment criteria can enable hospitals to generate revenues and save on equipment investment, peripheral consumptive materials, personnel costs and utilities.

6.領域的貢獻：你的提案對相關工作領域的貢獻？

範例 i.

The proposed fuzzy theory-based AHP method can enable local enterprises to evaluate the effectiveness of logistics centers based on a criterion index, subsequently increasing corporate competitiveness and efficiency by 25% while reducing personnel costs by 10%. Additionally, the proposed method provides an effective means of assessing the quality of services provided by logistics centers, Logistics centers must respond to specialized consumer demands.

範例 ii.

In addition to avoiding unnecessary risks and reducing overhead costs, the proposed decision-making method can incorporate an AHP evaluation model and several evaluation criteria for large-scale hospitals outsourcing their technological needs.

範例 iii.

Decision making based on the AHP criteria can provide a valuable reference for public hospitals that seek outsourcing partners. In addition to enabling Department of Health authorities to understand the full implications of outsourcing, the AHP-based assessment criteria can help public hospitals to understand the nature of problems that arise during outsourcing when constructing a deluxe health examination center.

範例 iv.

The proposed AHP-based method and fuzzy theory provides an effective means of assessing service quality provided by logistics centers, thus offering enterprise managers with the ability to enhance service quality and improve the strategy-tactics model for the logistics centers' strategic management procedures, ultimately enhancing the global competitiveness of Taiwanese enterprises by providing them with clear management guidelines for evaluating their suppliers.

7.摘要撰寫參考範例：

Setting of work proposal 工作提案建構 Banks loan to or invest in organizations based on whether they have a good credit rating, which involves investigations undertaken by financial experts examining financial reports to determine current status and classify credit results. Work problem 工作問題 Such time consuming and prohibitively expensive investigations do not always yield satisfactory results and even miss market opportunities. Work objective 工作題案目標Therefore, this work presents a novel classification model for small- and medium-sized enterprises, capable of increasing the prediction accuracy of classifying the credit rating of enterprises and reducing administrative expenses. Administrative expenses generally include employing financial experts, reducing time spent in examining credit ratings and handling wrong decisions. Methodology to achieve objective 達成目標的方法Data from small- and medium-sized enterprises are accumulated and categorized, followed by processing with a fuzzy set to store additional information. Fuzzy data are then clustered using a two-stage clustering approach to obtain a proper classification. Next an ANN structure is constructed by using the cluster results as an output of ANN. Additionally, the ANN classification machine is estimated with 10-fold cross validation to identify the most efficient machine. Moreover, results obtained from the above tests are tabulated and compared with those in literature to verify data authenticity. Main results of the project 希望的結果The proposed classification model can increase prediction accuracy from 10% (as achieved by the conventional classification system) to 80%, significantly reducing operational costs. Contribution to field領域的貢獻 In addition to increasing the efficiency of obtaining data from small- and medium-sized enterprises, and reducing operational costs during estimation processing, the proposed model provides promising guidelines for banks.

F. 工作提案撰寫（750-1000 words），由以上規劃，撰寫750至1000字的工作提案，包括五個段落：

第一段落：Setting of work proposal / 工作提案建構

參考範例：

Capability maturity model integrated (CMMI) consists of five maturity levels related to software development. Level 1 organizations have few or no rules for the software process. Level 2 organizations are better managed and the process is reproducible. Level 3 organizations adopt the same process applied to the entire organization. Level 4 organizations achieve quality management through statistical methods to manage projects quantitatively. Level 5 organizations have

continuously improving processes extended from Level 4 capabilities.
（請由以上主題延伸多寫5至7個英文句子）

第二段落之一：Work problem／工作問題

參考範例：

Although commonly adopted in manufacturing, statistical process control (SPC) methods such as control charts and process capability analysis are seldom used in the software industry. Despite the increasing popularity of using statistical tools in software development, some features in the software industry markedly differ from those in a manufacturing environment.
（請由以上主題延伸多寫5至7個英文句子）

第二段落之二：Quantitative specification of problem／問題的量化

參考範例：

For instance, software development activities are generally more process-based than product-based, making it difficult to apply control charts in a straight forward manner, especially for CMMI Level-4 organizations. Additionally, manufacturing and software processes significantly differ in that the latter is more human-intensive and creative. Moreover, according to Industrial Development Bureau data, no Taiwanese organizations or enterprises have passed the CMMI-Level 4 assessment.
（請由以上主題延伸多寫5至7個英文句子）

第三段落：Importance of problem／問題的中心

參考範例：

Although statistical approaches have been applied for software development, including monitoring & control, peer review and test processes, not all metrics are appropriate for SPC. Therefore, the inability to select appropriate metrics in software processes will hinder successful implementation of the SPC method in organizations, ultimately inhibiting the maturity of software processes.

（請由以上主題延伸多寫5至7個英文句子）

第四段落之一：Work objective／工作目標

參考範例：

Based on the above, we should develop a software metrics model that complies with CMMI-Level requirements, capable of considering how manufacturing and software process characteristics differ. Such a capability facilitates use of the SPC method in reducing scrap materials and work redundancy, increasing productivity and enhancing software quality. Given the well established role of control charts in SPC for identifying potential manufacturing problems, essential metrics for software process must be constructed to correspond with control charts.

第四段落之二：Methodology to achieve objective／達成目標的方法

參考範例：

To do so, a multiple objective decision method can be developed for selecting software metrics. A comparison between metric and CMMI Level-4 can then be considered as a criterion to screen the metrics. Next, generating capability of data for software development can be simulated to replicate the software process during life cycle. Additionally, properties involving variety of control charts (e.g., -R Chart, -S Chart, P Chart, NP Chart, C Chart and U Charts) can be analyzed on all estimated metrics. Moreover a case study can demonstrate the effectiveness of the proposed model.

第五段落之一：Anticipated results / 希望的結果

參考範例：

As anticipated, the proposed software metric method can contribute to efforts to increase the effectiveness of SPC in the software industry, especially with respect to analyzing defect density, percentage of work redundancy and inspection performance metrics. In addition to suggesting appropriate metrics for the software industry, the proposed method can ensure the effectiveness of using SPC, making it possible to analyze and transform several non-normal distributed metrics to comply with conventional control charts.

（請由以上主題延伸多寫5至7個英文句子）

第五段落之二：Contribution to field / 領域的貢獻

參考範例：

The proposed method also provides practical insight into the effectiveness of SPC for selected metrics in software development, as well as illustrates the merits and limitations of applying SPC to CMMI Level-4 compliant organizations compliant with CMMI Level-4. Importantly, the proposed model can assist Taiwanese organizations and enterprises in maturing software processes and enhancing software quality. Furthermore, the proposed method is also highly applicable to other non-manufacturing sectors, e.g. service or finance sectors.

（請由以上主題延伸多寫5至7個英文句子）

【參考文獻】

柯泰德（2002）。《有效撰寫英文工作提案》。台北：揚智。

Unit Seven

Writing Effective Academic or Professional Training Proposals

管理師學術及專業訓練英文申請撰寫

簡 介

本單元主要探討管理師如何撰寫學術及專業訓練英文申請，內容共分為六個部分，分別是：A.表達學習領域興趣；B.展現已有的學習領域知識；C.描述和學術及專業訓練相關的學歷背景及已獲成就；D.介紹學術及專業經驗；E.描述與學術及專業訓練有關的課外活動。管理師們可以經由參考研究他人寫作範例及相關輔助的九個練習，逐步強化本身對於學術及專業訓練英文申請的寫作能力，進而有效的提升競爭力。

A.Expressing interest in a field of study / 表達學習領域興趣

1. Stating how long one has been interested in a particular field or topic / 描述專業興趣所延續的時間

範例 i.

An article I read in the International Journal of Management during my junior year in university on the latest developments in strategic management aroused my interest in this rapidly developing area, eventually leading to my pursuit and successful completion of a post-graduate degree in this field.

範例 ii.

A book by a leading authority on e-commerce aroused my curiosity during university to obtain more information about management science.

範例 iii.

Quality management has fascinated me ever since I participated in a university-sponsored summer camp. My interest eventually led to my decision to acquire a bachelor's degree in this discipline.

範例 vi.

My interest in business administration stems from my conviction developed as an undergraduate student that economic development, while improving our lives, should adopt as many sustainable development practices as possible.

範例 v.

My fascination with management science can be traced back to adolescence when I often attempted to understand complex phenomena commonly found in commerce.

第一個練習：

以下每句爲用來訓練有關管理師學術及專業訓練英文申請撰寫：描述專業興趣所延續的時間。請把左邊及右邊的每一片語連結成一個完整句子。解答在本章最後。

1. Particularly fond of quality management curricula, I have always	a. Bachelor's degree in Business Administration from National Taiwan Ocean University.
2. While studying Management Science as an undergraduate student at National Taiwan University of Science and Technology,	b. introductory course that I enrolled in during university.
3. Strategic management intrigued me while I was	c. enjoyed researching related topics.
4. Having dreamed of becoming an entrepeneur since childhood, I acquired a	d. at National Taiwan University of Science and Technology with a concentration in sustainable development-related topics.
5. Individuals aspire to attend university in order to broaden	e. years at National Chiao Tung University, where I focused on environmentally friendly practices in technology management.
6. My fascination with business management started in an	f. pursuing a bachelor's degree in the Department of Business Administration at National Central University.
7. Having immersed myself in the accounting profession for quite	g. some time, I am currently pursuing a master's degree in this field.
8. My interest in researching quality management-related topics stems from my undergraduate	h. their horizons and pursue professional interests, explaining my interest in education from an early age.

9. My fascination with how commerce impacts all aspects of urban life explains why I am pursuing a Master's degree in Business Administration	i. interact with the younger generation.
10. The teaching profession has strongly interested me since childhood because educators can closely	j. I became intrigued with the role of sustainable development practices in management science.

2. Describing the relevance of one's interest to industry or society / 描述興趣與產業及社會的相關性

範例 i.

My interest in sustainable development-related practices in management science stems from its potential for significantly impacting our lives in unimaginable ways.

範例 ii.

In the information era, learning is a lifelong adventure that allows individuals continuously to upgrade their knowledge skills, to remain competitive in the tight labor market and avoid redundancy in the workplace.

範例 iii.

With finance among the fastest growing sectors in newly industrialized countries, Taiwan is widely anticipated to become a hub of development in this rapidly evolving area.

範例 vi.

The Taiwanese government has restructured local financial markets in recent years. Numerous opportunities for growth expansion in the finance field explain my interest in investment within the hi-tech sector.

範例 v.

With Taiwan's recent accession to the World Trade Organization, solid design and manufacturing capability aside, as well as financial expertise in global markets, will be a necessity for Taiwanese companies. I am especially interested in the financial aspects of how local hi-tech firms will become globally competitive.

3. Stating how one has pursued that interest until now / 描述興趣形成過程

範例 i.

A solid academic background and abundant research experience in business management make me confident of becoming a competent manager. I believe that these two ingredients are crucial to my fully realizing my career aspirations.

範例 ii.

The strong departmental curricula at National Chiao Tung University instilled in me academic fundamentals necessary for advanced research. Financial Management captivated my interest because of its complexity and diverse applications.

範例 iii.

My dream of becoming a widely respected manager who will design innovative programs has motivated me to pursue a PhD. in Technology Management at National Taiwan University in preparation for a career in research.

範例 iv.

Success in my field of interest depends on developing a truly multidisciplinary approach, which I tried to nurture during undergraduate and graduate studies by exposing myself to a diverse array of seemingly polar topics and searching for their possible relationships in a practical context.

4. Stating how academic study or professional training related to one's interest would benefit the applicant / 描述學術及專業訓練和興趣配合的利益

範例 i.

The stringent six-month training program offered by your company will equip me with the skills to thrive in the civil engineering profession and to more significantly contribute to my consultancy firm's offered services.

範例 ii.

Your excellent research environment and abundant academic resources will enhance my research capabilities and equip me with the competence to fully realize fully my career aspirations.

範例 iii.

The opportunity to receive training at your company would allow me to fully realize my career aspirations. The quality control group in your company is the undisputed leader in this field in Taiwan. Renowned for its strong organizational culture and management structure, your company offers professional training that can enhance my work.

範例 iv.

Interacting with the highly skilled professionals of your company will undoubtedly provide me with a practical context for the academic training that I have received.

範例 v.

I am impressed with your company's commitment to nurturing professionals in the finance sector through your comprehensive six-month training program, from which I will definitely benefit.

第二個練習：

以下每句爲用來訓練有關管理師學術及專業訓練英文申請撰寫：描述學術及專業訓練和興趣配合的利益。請把左邊及右邊的每一片語連結成一個完整句子。解答在本章最後。

1. Exposure to the marketing practices and data analysis procedures adopted in your	a. implement effective management strategies for the finance sector.
2. The opportunity to work in a practicum internship in your company would provide me with an excellent environment not only to fully realize my career aspirations,	b. for my lack of formal professional experience in this area.
3. As a leader in this field of research, your laboratory could provide	c. consultancy firm would fully orient me on the latest trends in management science.
4. The opportunity to receive technical training in your company’s research unit would equip me with the skills necessary to design and	d. me a clearer direction of not only my graduate level research, but also my chosen career path as well.

5. The opportunity to participate in your research program would pave	e. I would receive a practical context for the solid academic fundamentals acquired so far.
6. Your organization has much to offer management	f. but also to implement quality management methods in the banking sector efficiently.
7. The opportunity to receive technical training in your organization would compensate	g. like to serve in a self-supported practicum internship in your organization during my upcoming summer vacation.
8. If given the opportunity to serve as a visiting researcher in your laboratory,	h.me with the exposure to the latest trends in finance from a global perspective.
9. Given your consultancy firm's widely reputed image as adopting the latest approaches to comply with industrial requirements in the finance sector, I would	i. the way for collaborative activities between our two laboratories, based on our mutual interests.
10. Exposure to the way in which your department applies advanced engineering concepts in a practical context would give	j. professionals in a newly industrialized country such as Taiwan.

5.Samples of expressing interest in a field of study

While pursuing a Bachelor's degree in Information Management at Nanhua University, I was exposed to a wide array of subjects, including management and business-related courses. Although a university instructor encouraged me to major in Information Management given its popularity and innovative nature, many other universities were producing graduates from this discipline, making it fiercely competitive to gain employment in this line of work. This situation motivated me to acquire other skills in order to become more competitive in the workplace and have more job prospects. This explains why I am currently pursuing a Master's degree in Business Management at Yuanpei University. I am especially interested in finance, which I believe will complement my background in information management. As a more established discipline with many potential areas of development, finance is a necessity for individuals or enterprises that requires long term planning. My strong interest in finance arises from two sources. First, my high marks in economics and financial management courses during university made me realize the strong financial skills that students can acquire, which will greatly enhance their knowledge skills and ultimately lead to higher productivity. Second, individuals need greater knowledge skills or higher forms of entertainment, all of which require money. Some students face large debts following university graduation owing to increasingly higher tuition costs and the inability of their families to support them financially. Meanwhile, other individuals simply can't manage their finances, explaining why they have no or little savings and must spend their income to pay for accumulated debt. Therefore, more than just earning a lot of money, finance is also learning how to cope with outstanding debts. Perhaps the dream for most individuals is not to worry about finances.

Mathematics has enthralled me since childhood, explaining why I majored in Finance and Banking during undergraduate school after passing a highly competitive nationwide university entrance examination. Strengthening my analytical skills enables me not only to analyze economic situations precisely, but also to evaluate financial data of a company. After receiving my bachelor's degree from Tamkang University, I obtained six government-approved licenses that not only qualified me to work in Taiwan's financial sector, but also enabled me to easily secure employment. When the opportunity arose to pursue a Master' s degree in Business Management at National Central University (NCU), I decided to continue with my studies following careful consideration. Coming to graduate school, I have already acquired five years of work experience given that I began my undergraduate studies after completing our country's compulsory military service. As my work has fortunately been related to the finance and banking sectors, I gained much experience in banking practices, financial derivatives for trading purposes and analysis of financial reports for stock market and futures trading. Hopefully, this exposure to financial management will prove beneficial to my graduate studies. Not necessarily intellectual, I consider myself diligent in grasping academic concepts and applying them in a practical context. I am thus confident that NCU can equip me with the advanced skills necessary to thrive in my specialized field. Expecting to fulfill my graduate studies at NCU in two years, I hope to secure employment as a finance-related researcher, economic analyst or financial management professional. Doing so would enable me to apply the concepts taught in graduate school to the workplace. My teacher's recommendation of NCU and the graduate school's emphasis on strengthening English language skills for research publication and professional purposes further reinforced my decision to pursue advanced study here.

Having recently graduated from the Department of Finance at Taichung Healthcare and Management University, I studied diligently during my junior and senior years. Given my strong interest in continuously upgrading my knowledge skills, I gained entry into the Graduate Institute of Business Management at Yuanpei University in pursuit of a master's degree. As an increasing number of Taiwanese universities offer management school programs, I would like to acquire more management-related knowledge, such as in marketing, statistics for business purposes and economics. Moreover, as English has become the medium of communication worldwide, I must enhance my listening, speaking, reading and writing skills to remain competitive in the workplace. Despite the above challenges, I am confident of success in my academic studies as long as I closely follow my teachers' instruction. In sum, successful academic study and high marks require that a student goes the extra mile. With my ultimate goal of working in the stock and bond markets, I must acquire governmental certification to become qualified for work in the finance sector in addition to my graduate level education. Despite the widely held notion that professionals in the stock and bond markets are greedy and predatory, professional ethics is essential to survive in this profession. Doing so requires stamina and a perseverant will.

My interest in marketing stems from my background in commerce, as my family has set the example for working diligently ever since my childhood to ensure that I would have a higher living standard. Understanding consumer demand in design is a form of exploratory psychology. A product may have added value given the marketing attention or strategy that is adopted. A product purchase not only satisfies consumer demand, but also increases the value of essential items that are purchased. In a competitive market, promotion of a reliable product or service quickly leads to imitation from competitors, necessitating that the original manufacturer strengthens its product superiority to maintain consumer loyalty. Modern society heavily emphasizes product performance to acquire an advantageous position in a particular market niche. Doing so requires an emphasis on performance, logistics flow and customer relations. I am especially intrigued with this research topic. When studying marketing approaches, I closely scrutinize the current market situation from a management perspective. As a junior high school student, I easily obtained a license to work as a chef. During that period, I was strongly interested in marketing and decided to pursue advanced study. While working for the Wang Steak Group, I adopted four marketing strategies to enhance the storage management capacity of the restaurant. I was able to coordinate all purchases of food ordering, which gave a clear sense of procurement control. Following university graduation, I worked in Powerchip Semiconductor Corporation as a technician in on-line production. This experience made me aware of the complexity that the management of a large organization faces, motivating me to research management-related topics in graduate school.

As a junior high school student, I was not interested in the academic curricula of senior high school, explaining why I studied in a vocational school for commerce instead. During this period, I became interested in accounting and economics. This was my main motivation for majoring in Healthcare Management at Yuanpei University (YPU). While gaining a firm understanding of management practices in Taiwan's medical treatment facilities, I received specialized training in statistics medical information systems and ASP.net. My senior thesis prior to graduation focused on the effective management of hospital resources by using statistical methods and the gray mathematical theory. Using these methods, I attempted to determine the amount of hospital resources used, such as accurately forecasting the amount of intravenous injections in a hospital. Following university graduation, I decided to continually refine my knowledge skills in medical management by pursuing a Master's degree in the Institute of Business and Management at the same university. Given my general interest in healthcare management, I hope that graduate school will enable me to specialize in areas such as financial management and marketing. Becoming knowledgeable of the latest trends in finance would enable me to actively participate in financial management research, perhaps related to capital funds and the stock market. While offering indispensable knowledge, financial management can orient individuals on how not only to manage capital efficiently, but also to acquire basic commercial skills. As an integral part of business, marketing significantly influences the commercial success of products. An effective marketing strategy could encompass the latest technology skills, advertising or novel sales concepts, thus ensuring corporate survival in a highly competitive environment. The marketing-oriented curricula in the Institute of Business and Management at YPU will strengthen my professional knowledge skills.

B.Describing the field or industry to which one's interest belongs／展現已有的學習領域知識

1. Introducing a topic relevant to one's field of study／介紹和學習領域相關的主題

範例 i.

Despite political differences, China and Taiwan share many common economic interests, as evidenced by their increasingly strong economic ties in the Asia-Pacific region. I am interested in understanding how the finance sector can contribute further to strengthening these ties.

範例 ii.

The explosive pace at which strategic management practices appear in the workplace never ceases to amaze me, and I am eager to remain abreast of the latest developments in the field.

範例 iii.

Given the island's recent accession to the World Trade Organization and the eradication of many trade barriers to the local market economy, local entrepreneurs must continually upgrade their professional skills and linguistic abilities to remain competitive in the global workplace.

範例 iv.

Taiwan severely lacks technology professionals with a background in engineering and business administration, preventing the island's economy from keeping up with recently emerging trends in global R&D.

範例 v.

Following its rapid industrial development in recent decades, Taiwan is now striving to become the 'Green Silicon Island' by simultaneously pursuing environmentalism and developing high-tech industries, areas that quality management professionals are committed.

第三個練習：

以下每句為用來訓練有關管理師學術及專業訓練英文申請撰寫：介紹和學習領域相關的主題。請把左邊及右邊的每一片語連結成一個完整句子。解答在本章最後。

1. With its rapid industrial development in recent decades, Taiwan is striving to become	a. to the potential for integrating seemingly polar disciplines in order to create new technical and market opportunities.
2. Taiwan severely lacks technology professionals with a background in engineering and business administration,	b. design various products to ensure that our lives comfortable and convenient.
3. Pursuing economic growth while protecting the environment is a careful	c. preventing the island economy from keeping pace with this recently emerging trend in global R&D.
4. Quality management professionals analyze processes and	d. despite a global recession, has recently entered the World Trade Organization and prepares to host the 2008 Olympics.

5. The finance sector in Taiwan is rapidly expanding though,	e. the 'Green Silicon Island' by simultaneously pursuing environmentalism and developing high-tech industries.
6. Integrating environmentally friendly practices with civil engineering attests	f. unfortunately, this expansion is largely in the quantity of work produced, rather than the quality.
7. The global focus on commercial opportunities in China is intensifying as the country, in addition to achieving an annual economic growth of 7%	g. balancing act performed by most industrialized economies.
8. Although Taiwan's hardware industry has already matured	h. its software industry lags behind that of many other developing economies.
9. Recently, Taiwan has significantly contributed to the global high-tech sector,	i. the island economy's investment strategies and enhance global competitiveness.
10. Financial planners in Taiwan constantly focus on how to adjust	j. especially in microchip fabrication, electronic packaging and IC design.

2. Describing the importance of the topic within one's field of study / 強調學習領域主題的重要性

範例 i.

Integrating information technology and quality management into the burgeoning finance sector in Taiwan has tremendous commercial and societal implications for the island's economy.

範例 ii.

Adopting knowledge management practices in the banking sector attests to the potential for integrating seemingly polar disciplines to create technical and market opportunities.

範例 iii.

Taiwan's pivotal role in contributing to the infrastructure development of Greater China necessitates that local management professionals and investors remain attuned to the latest technological and developmental trends on both sides of the Taiwan Strait.

3. Complementing the organization to receive training from on its commitment to excellence in this area of expertise／讚美所給予專業訓練的機構

範例 i.

The great working environment, combined with the impressive number and diversity of training courses to maintain the competitiveness of your employees in the market place, would definitely benefit my professional development.

範例 ii.

The diversity of research projects and departments within your company committed to implementing them is quite impressive.

範例 iii.

As a leading consultancy firm, your organization has distinguished itself in its commitment to in-house research projects, as evidenced by your excellence in product innovation.

範例 iv.

After reviewing your online brochure and research findings published in international journals, I am quite impressed with your laboratory's facilities and the wide scope of activities that you are engaged.

範例 v.

After thoroughly reviewing your online literature and promotional materials, I am especially drawn to the marketing strategies that your organization has adopted, which are quite compatible with my current research interests.

第四個練習：

以下每句為用來訓練有關管理師學術及專業訓練英文申請撰寫：讚美所給予專業訓練的機構。請把左邊及右邊的每一片語連結成一個完整句子。解答在本章最後。

1. As your laboratory has devoted much time in researching the above topics, I would greatly benefit from	a. your findings published in international journals.
2. As a leader in this field of research,	b. quality management research for quite some time, developments that I have closely followed for quite some time.
3. While searching for pertinent research literature in my field, I often come across	c. as evidenced by the publishing of your research findings in several internationally renowned journals.
4. As is well known, your research group has	d. the most appropriate place.
5. From your laboratory's online introduction, I am impressed with	e. as well as its globally recognized services for improving the lives of urban dwellers.
6. Your company has distinguished itself in	f. in researching quality management-related topics.

7. As is well known, your research group has made considerable progress in adopting sustainable development in management practices,	g. the opportunity to work in your laboratory as a self-supported researcher during my upcoming summer vacation.
8. Your division is widely recognized as a leader	h. the comprehensive strategies that you adopt on the work site.
9. Given my research interests, your laboratory appears to be	i. your laboratory could allow me to further refine my civil engineering-related skills.
10. I hope to perform advanced research in your center owing to its highly skilled personnel, state-of-the-art equipment and facilities,	j. devoted considerable time and resources to researching construction design-related topics.

4.Samples of describing the field or industry to which one’s interest belongs

Practiced worldwide, finance ranges from banking, stocks and other investments. Of great concern to all individuals, finance in Taiwan is especially important in banking, since many employers remit staff salaries through banks; Chinese also generally save a relatively high proportion of their income. As an increasing number of businesses change their commercial practices, individuals rely more on their banks than previously. Finance is thus an enduring discipline that can not be replaced, regardless of the many practices that it encompasses. In addition to banking, the insurance sector has emerged in Taiwan in recent years as an integral part of finance. More than investment, insurance protects individuals from many unforeseeable crises in life. Without it, many individuals would not have the resources to protect their families. Restated, finance comes in many forms, and individuals must learn how to remain as informed as possible. Simple financial practices such as saving and borrowing are important to all at some point in time.

While humans must purchase necessities such as food, water and shelter for survival, many individuals lack the financial resources to purchase basic essentials let alone luxury items. At this time, they must borrow money from banks to purchase those items, leading to acquired debt. However, individuals with more money than what their needs demand can save that amount in banks for the future. As an incentive, banks offer interest rates for individuals saving money in their care. Given the many forms of acquired debt and interest incentives, individuals are encouraged to spend or save more than previously. For instance, students faced with higher tuition costs and daily living expenses, but no income, fully realize the importance of banks in achieving their professional goals. In sum, given the many facets of finance in society, individuals must remain as informed as possible to take advantage of what is available.

Financial institutions can be categorized as banks, insurance companies, securities firms and investment banks. While banking focuses mainly on handling the debit and credit aspects of customer service, insurance companies sale insurance policies for personal or corporate protection. Additionally, while securities firms purchase and sale securities, investment banks have the widest scope of commercial activities among all financial institutions, with activities including underwriting, company mergers and business evaluations. Taiwan' s entry into the World Trade Organization has heightened the importance of financial institutions in recent years. Financial reforms in these institutions can especially be found in governmental policies and manufacturing practices. The WTO' s significant impact on financial institution has led to a dramatic shift not only in domestic financial resources, but also in consumer demand on the overall financial system. Simultaneously, WTO' s impact has led to tremendous financial reform among local institutions. Among those reforms included a decrease in poor performing financial institutions and the integration of smaller ones into larger, fiscally sound organizations. Such measures have attempted to increase the island's global competitiveness in the financial sector. Financial holding companies have thus evolved through the continuous integration of financial institutions. The Taiwanese Government has aggressively adopted financial reform strategies in recent years, with intentions of transforming the local financial sector into a more professional and diversified global competitor that can comply with international trends. However, financial reform has encountered some obstacles. For instance, governmental intervention has failed to instigate progress within financial institutions as quickly as originally anticipated. Additionally, stringent overseas competition has somewhat stymied the progress of financial reforms.

Implementation of the National Health Insurance scheme in Taiwan has gradually led to an overall decline in hospital revenues. As a high overhead cost organization, hospitals spend much capital on human resources and pharmaceuticals. Therefore, efficient cost management with an emphasis on medical expenditures and cost control is essential to resolving capital-related problems. Overhead costs in a hospital can be divided into four constructs: distribution, stock, operation and procurement. Viable options to reducing overhead costs include distributing materials to a specific location, outsourcing inventory management, operating through electronic commerce and purchasing through joint procurement. Therefore, hospitals heavily concentrate on simplifying administrative procedures and facilitating professionals to enhance service quality. Retailers such as 7-11 Convenience Store have recently implemented the third party logistics service provider (3PL) model to distribute goods to a specific location. Above measures should enable the medical sector to reduce unnecessary costs, create added-on value and replace ineffective management methods with logistic services that integrate logistics, information management and automated data links. My graduate school research will focus on analyzing regional hospitals and medical centers through expert interviews to identify those factors that enable hospitals to implement 3PL in pharmaceuticals. In addition to a hospital's emphasis on performance, deployment and inventory management, my research will incorporate 3PL's emphasis on providing, exchanging and utilizing information to explore determinants of performance with respect to the status of information and relationships among organizations. Results of my research will ultimately provide a valuable reference for hospital administrators in preliminary planning and project implementation.

In addition to initiating aggressive financial reforms, the Taiwanese Government has initiated changes in the island's monetary system. The local banking sector has faced numerous challenges in recent years. The monetary system has played a prominent role in Taiwan's economic development over the past five decades. However, when a financial crisis arises, banks are negatively impacted as well. Despite the resolve of Taiwan's president to resolve instability in the financial system, many problems still persist. Irregularities in bank loans are increasingly serious, leading to a crisis within banking institutions. For instance, stock prices in Taiwan were low, motivating the government to adopt remedial measures to stabilize the stock market. A depressed economy creates greater pressures in daily life. Despite these circumstances, Taiwan should strengthen the global competitiveness of its financial institutions. Meanwhile, the Taiwanese Government strives to elevate the living standard of its residents. Many people believe that employment in the financial sector can create more stability and higher salaries than they could find in other fields. Governmental reforms in the financial sector will lead to a reduction in the number of banking institutions island wide and, ultimately, increased economic benefits through the consolidation of these institutions. In sum, integration of economic resources in the banking sector will ultimately spell success in local financial reform.

Initiated in 1995 to provide comprehensive health care, the National Health Insurance (NHI) scheme in Taiwan provides comprehensive coverage for all island residents. Given its significant contribution to the health service sector, the NHI Bureau has gradually implemented a case payment scheme and Global Budget system for dental, primary and hospital care in recent years. As the NHI payment system continuously evolves, hospital administrators strive to cope with policy changes as effectively as possible. Additionally, health care expenditures have rapidly increased, with outpatient services accounting for up to 70% of the total. Therefore, the co-payment scheme simultaneously attempts to control consumer demand and prevent the abuse of health care resources. Given my background in healthcare management, I am interested in market trends in this sector and current management trends. However, since the NHI was initiated, Taiwan's demography and living standards have gradually changed. The leading causes of mortality have switched from acute infectious diseases to chronic ones, with diabetes a notable example. Diabetics are largely middle aged or elderly individuals. With its high incidence rate, diabetes-related complications and mortality consume a significant amount of medical resources. Given the growing constraints on hospital revenues and quality care of chronic illnesses such as diabetes, cost control and management should be of priority concern. Hopefully, my graduate school research will discuss how financial factors arising from NHI policies and subsidy payment systems affect treatment of diabetes within the context of the healthcare market and hospital management practices involving the subsidy payment system of NHI.

C. Describing academic background and achievements relevant to academic or professional training／描述和學術及專業訓練相關的學歷背景及已獲成就

1. Summarizing one's educational attainment／總括個人的學術成就

範例 i.

I majored in Business Administration during university and later received a master's degree in the same field, with the specific intention of acquiring sufficient expertise to enable me to contribute effectively to society in the future.

範例 ii.

Immediately following four years of highly theoretical study and intellectual rigor at National Taiwan University in successful pursuit of a bachelor's degree in Finance, I gained entry to the highly prestigious doctoral program in the same field at National Cheng Kung University.

範例 iii.

I received my undergraduate and graduate training in Financial Management at National Chiao Tung University and National Tsing Hua University, respectively. These are both higher institutions of learning at which many staff at the Hsinchu Science-based Industrial Park developed their professional skills.

範例 iv.

While my Master's degree in Management Science prepared me for the rigorous demands of conducting original research in this field and publication of those findings in international journals, doctoral study in the same field equipped me with the required knowledge skills and professional expertise to excel in the quality management profession.

範例 v.

Undergraduate studies exposed me to diverse disciplines, thus allowing me to choose a career path more objectively than if I had a bias for a particular area.

2.Describing knowledge, skills and/or leadership qualities gained through academic training／描述學術訓練所獲得的技能及（或）領導特質，請參考以下範例：

- Critical thinking skills developed during undergraduate and graduate studies have enabled me not only to explore beyond the first appearances of finance-related issues and delve into their underlying implications, but also to conceptualize problems in different ways.

- In the accompanying academic transcripts, high marks in those courses that equipped me with a solid theoretical background in business administration reflect my commitment to acquiring knowledge and skills that will not only make me competent, but also allow me to thrive in the workplace.

- Undergraduate and graduate level courses in management science often involved term projects that allowed me to apply theoretical concepts in a practical context and to develop my problem-solving skills.

- My diverse academic interests reflect my ability to see beyond the conventional limits of a discipline and fully comprehend how the field relates to other fields. Solid training at this prestigious institute of higher learning equipped me with strong analytical skills and research fundamentals.

第五個練習：

以下每句爲用來訓練有關管理師學術及專業訓練英文申請撰寫：描述學術訓練所獲得的技能及（或）領導特質。請把左邊及右邊的每一片語連結成一個完整句子。解答在本章最後。

1. My diverse academic interests reflect my ability to view	a. resolve to apply my newly acquired knowledge and skills in an interdisciplinary manner.
2. Undergraduate and graduate level courses in business management often involved	b. and developed my ability to work independently.
3. Through my academic training, I acquired specialized knowledge of finance-related issues	c. aware of the importance of financial planning and evaluating a project's success.
4. Equipped with strong analytical and experimental skills acquired during graduate school,	d. in undergraduate management courses facilitated my understanding of how this field can be integrated in the hi tech sector.
5. Besides developing my academic abilities, the strong departmental curricula strengthened my	e. beyond the conventional limits of a discipline and fully comprehend how it relates to other ones.
6. Graduate study in Business Management made me	f. as evidenced by my doctoral dissertation, and some of those findings were published in several international journals.
7. A rich blend of theoretical knowledge and practical concepts taught	g. I am able to find a practical context for applying the theoretical concepts taught in the classroom and understand how they relate to the hi tech sector.
8. Doctoral studies increased the breadth and depth of my understanding of strategic management,	h. term projects that allowed me to apply theoretical concepts in a practical context and to develop my problem-solving skills.

9. By equipping me with the fundamentals of computer science,	i. but also allow me to thrive in the workplace.
10. In the accompanying academic transcripts, high marks on those courses that equipped me with a solid theoretical background in business administration reflect my commitment to acquiring knowledge and skills that will not only make me competent,	j. undergraduate courses underpinned the professional training that I received in graduate school.

3. Emphasizing a highlight of academic training／強調一個學術訓練的特定領域，請參考以下範例：

範例 i.

My graduate school research often involved deriving complex mathematical models and then presenting those results clearly, both in writing and orally. This continual activity of explaining concepts to those outside of my field of expertise was invaluable training to meet a need that I often encounter in the workplace — to provide sufficient details to non-technical professionals to enable them to make management decisions.

範例 ii.

Graduate school instilled in me the need for harmony among research collaborators to achieve a desired outcome. As a group leader, I often organized discussion groups on how to implement operating systems developed in our laboratory. In addition to exchanging relevant theoretical knowledge and individual expertise, our group formed bonds that greatly facilitated accumulation and analysis of data and, eventually, the publishing of our research findings in several renowned international journals.

範例 iii.

My curiosity and determination to adopt unconventional or multi-disciplinary approaches to solve quality management problems not only distinguished me from other graduate research students, but also led to my developing a management consultancy program that has already been commercialized.

範例 iv.

While heavily emphasizing self-learning and development aimed at contributing to collaborative research efforts, graduate school ignited within me a strong desire to diversify my knowledge and skills, and thus adopt multi-disciplinary approaches in an industrial setting.

範例 v.

Studying at National Cheng Kung University, one of Taiwan's premier institutes of higher learning, strongly motivated me to acquire the latest knowledge skills and technical information. More specifically, I learned how to analyze problems, find solutions and implement those solutions according to the concepts taught in class.

範例 vi.

I also participated in a research project during my senior year, learning how to be a contributing member of a team and develop a thorough understanding of teamwork. I also attended several international conferences that addressed quality management-related issues to broaden my perspective on potential applications of management and information technology— these areas are now my main areas of interest.

第六個練習：

以下每句為用來訓練管理師學術及專業訓練英文申請撰寫：強調一個學術訓練的特定領域。請把左邊及右邊的每一片語連結成一個完整句子。解答在本章最後。

1. My curiosity and determination to adopt unconventional or multi-disciplinary approaches to solve networking problems not only distinguished me from other graduate research students,	a. on database applications in industry required me to collaborate closely with classmates by simulating an industrial project to fulfill course requirements.
2. A cumulative grade point average of 4.0 during graduate school attests	b. for harmony among research collaborators to achieve the desired outcome.
3. During university, I constantly read commentaries and papers on the latest developments in business administration to	c. to my professional grasp of the latest technological trends in management science.
4. Besides providing a practical context for the academic fundamentals of management science, a senior seminar	d. multi-disciplinary approach to integrating knowledge and skills from different fields.
5. A graduate seminar on Internet-based technologies made me	e. broaden my perspective on potential technological and industrial applications.
6. Writing my Master's thesis proved to be a valuable exercise in adopting a	f. but also led to my developing a software program that has already been commercialized.
7. Graduate school instilled in me the need	g. aware of their rapid growth and incorporation into areas in which they had not been previously applied, such as planning and design of project schedules.

8. This course motivated me to identify how information science concepts can applied in other disciplines to create potential	h. - the need to provide sufficient details for non-technical professionals to make management decisions.
9. Continuously explaining concepts to those outside my field of expertise was invaluable training for what I often encounter in the workplace	i. and then presenting the results clearly in written form and orally.
10. My graduate school research often involved deriving complex mathematical models	j. industrial applications that would have been otherwise impossible without such an integration of seemingly polar fields.

4. Samples of describing academic background and achievements relevant to academic or professional training

Although born in Yunlin, I grew up in Chaiyi, living there for fifteen years. I continued living in Chaiyi while pursuing a Bachelor's degree in Information Management from Nanhua University. Besides information management, my university instructors encouraged me to study economics, management and various computer software programs. I also learned much about computer hardware, including its components, repair and software applications. Before serving in Taiwan's compulsory military service for nearly two years, I spent much time in the Computer Center at National Chung Cheng University, where I learned a lot about Internet-based applications, such as operating computers in a wireless network environment. Although lacking formal work experience, I often worked part time during my undergraduate study by repairing computers and completing software programs. In sum, I have constantly strived to upgrade my knowledge skills, especially with respect to repairing computers and understanding software applications. Understanding computer hardware involves becoming fully oriented with not only the features and functions of computer components, but also hardware problems that arise from computer architecture, as well as operational knowledge of computer software and compilation of computer language programs. In addition to acquiring many computer-related knowledge skills, I enrolled in many undergraduate business courses, including Economics, Management and Accounting. I decided to compensate for my lack of advanced computer skills and management skills by pursuing a Master's degree in Business Management at Yuanpei University. Despite my above attempts to acquire such skills, graduate school is the next step for me to strengthen my financial management and investment capabilities.

While attending vocational high school in Miaoli of northern Taiwan, I concentrated on electrical engineering studies. During this period, I studied diligently and enthusiastically helped others when the opportunity arose. My efforts were recognized in certificates of commendation from both the city mayor and my teachers. Such recognition further encouraged me to apply myself as much as possible to academics. Upon graduation and following an intensely competitive nationwide university entrance examination, I gained entry to the Department of Biomedical Engineering at Yuanpei University. Undergraduate school instilled in me sound fundamentals of electronic devices and instrumentation for medical purposes. I especially became proficient in 8051 IC chip design, learning of the computer components involved in such machinery design. As a junior, my classmates elected me class leader, which made me more conscientious about the needs of others. I view myself as a highly responsible individual, as evidenced by the numerous tasks that classmates and teachers required of me. Despite serving as class leader for four semesters, I did not allow my academic studies to suffer, as reflected by a final academic ranking of second out of the entire class. On the contrary, those additional responsibilities offered valuable lessons in leadership that could only be learned through perseverance. My above academic experiences have prepared me well for the rigorous demands of pursuing a master's degree.

I received a Bachelor's degree in Banking and Finance from Tamkang University in 2005. While the undergraduate curricula provided me with solid fundamentals in banking and finance, I worked in the finance sector during this period as well to find a practical context for the academic concepts taught in class. In addition to learning about business mergers, financial statement analysis and acquisition methods in my academic studies, I also grasped the fundamentals of derivative commodities. To prepare for a career in this field, I obtained six government-approved licenses to practice in the finance sector. Despite my five years of professional experience, I still felt that my knowledge skills in the above areas were lacking. Therefore, following an intensely competitive nationwide graduate school entrance examination, I gained entry to the Institute of Business Management at Yuanpei University (YPU). My most recent employment involved serving as an assistant analyst at Jihsun Holding Company, which oriented me on futures trading, option transactions, marketing analysis capabilities and stock market principles. While strengthening my interests in a career in the finance sector, I hope to continuously upgrade my knowledge skills to forecast market trends in stocks, as well as the futures and options markets. However, doing so requires that I enhance my statistical and programming skills. Given the importance that licensed certification plays in working in the finance sector, I will continue to prepare for local and overseas accreditation examinations. With my current pursuit of a Master's degree in Business Management at YPU, I feel that my academic and professional experiences have aptly prepared me for study of finance at the graduate level to conduct independent research. I am especially determined to grasp advanced theoretical concepts in marketing, multivariate statistical analysis and other relevant research methods. More than facilitating my future research efforts, such skills will prove invaluable in resolving obstacles in the workplace, explaining why I am determined to take advantage of this opportunity in graduate school as much as possible.

As a native of Hsinchu, I graduated from a local high school with a concentration in general studies in 2001, followed by successful completion of a Bachelor's degree in Healthcare Management at Yuanpei University (YPU). Although the youngest child in my family, I learned how to become financially independent early on. Even throughout university, I worked part time. Still, I ranked among the top three academically in my class throughout school, reflecting that I hold myself to high standards. Additionally, I served in a practicum internship during my summer vacation of 2003 in an administrative department at Mackay Memorial Hospital in Hsinchu, Taiwan. This exposure enabled me to understand how a medical staff performs differently from employees in other organizations. A hospital focuses on the following areas: organizational structure, patients, the community that it serves and its staff. A healthy organization encourages the physical and mental health of its staff which, hopefully, is passed on to its patients and the surrounding community. During the internship, I attempted to develop an assessment procedure to evaluate the characteristics of a healthy medical organization in Taiwan. However, realizing my limited professional knowledge of healthcare management, I entered the Institute of Business Management at YPU in pursuit of a master's degree. My graduate school research focuses on how Taiwanese hospitals can operate under the National Health Insurance scheme while promulgating the Global Budget system under sound management strategies to avoid bankruptcy. I view myself as introverted and unable to fully express myself at times. As a university sophomore, I worked as a membership recruiter for JUSCO Corporation, learning how to communicate freely with customers and other employees. This invaluable experience enabled me to express myself without fear of others' reactions. As a graduate student, I am now able to freely express my opinions from a podium.

I received a Bachelor's degree in Finance from Yu Da College of Business in June of 2005.The department curricula provided me not only with theoretical concepts, but also a practical understanding of the finance sector, especially in areas of accounting, financial management, financial statement analysis, investments and civil law. I pursued my undergraduate degree at night so that I could work during the daytime. Even though night school was limited in time, resources and qualified teachers, I was determined to make the most of my academic study. During my junior and senior years, I secured employment as a bookkeeper in the finance and accounting department of an overseas trading company, which heightened my interest in finance-related topics. Coming into contact with actual enterprises allowed me to more fully understand the theoretical concepts taught in class. Although night school prevented me from directly working with my academic advisor on a research project, I obtained Techficiency Quotient Certification (TQC) in Web design. I even designed a website for our class that featured the photographs and short biographies of all of the classmates. Additionally, I served as an administrative assistant in the logistics department of an overseas trading company for three months. Following a transfer to the finance department, my supervisor oriented me on finance and accounting topics, which further heightened my interest in the finance sector. Majoring in Finance and working in an overseas trading company greatly enhanced my finance-related knowledge skills. Theoretical knowledge and practical experiences go hand in hand in terms of increasing my commitment to this field of study. Such experiences will definitely benefit my graduate school research.

D. Introducing research and professional experiences relevant to further academic or professional training / 介紹學術及專業經驗

1. Introducing one's position and/or job responsibilities, beginning with the earliest position, and ending with the most recent one / 介紹個人歷年至今所任工作職位及職責，請參考以下範例：

範例 i.

Intensive laboratory and theoretical training in a summer internship before my senior year in university motivated my decision to pursue a finance-related career and remain abreast of the latest developments in this rapidly evolving field.

範例 ii.

Serving as a research assistant in several National Science Council-sponsored projects in technology management familiarized me with the operations of a research laboratory, including weekly progress reports and regular seminars. I found the work highly stimulating and rewarding. Given my past academic performance and recent laboratory experience, I am confident of my ability to contribute to your company's R&D efforts.

範例 iii.

After serving for five years as an administrative assistant to the general manager, in which role I balanced between goals and departmental progress, I was promoted to assistant manager of the procurement department, becoming adept in gathering information, evaluating data, negotiating prices, and reviewing budgets.

範例 iv.

My enclosed curriculum vitae reflects extensive management experience in the quality management profession, acquired over more than a decade, made possible by graduate training at one of Taiwan's finest institutes of higher learning and by collaboration with many globally renowned industrial partners.

第七個練習：

以下每句爲用來訓練有關管理師學術及專業訓練英文申請撰寫：介紹個人歷年至今所任工作職位及職責。請把左邊及右邊的每一片語連結成一個完整句子。解答在本章最後。

1. While effectively coordinating personnel and other resources in the department, I was also responsible for	a. engaged in planning policy-making projects on various scales and with different aims.
2. As an administrative manager, besides fostering a closer relationship with our customers,	b. to pursue a career in finance to remain abreast of the latest developments in this rapidly evolving field.
3. As an administrative coordinator in an environmental protection agency, I actively	c. but also to coordinate such activities efficiently to maximize inter-departmental harmony and, ultimately, corporate profits.
4. I began working as a financial planner in a well-known	d. identifying market trends across the entire semiconductor industry and assessing our company's most effective responses.
5. Intensive laboratory and theoretical training in a summer internship before my senior year of university marked a turning point as I decided	e. my limited knowledge of how global economic activities affect small- to medium-sized firms.
6. My previous work experiences attest to my ability not onlyto comprehend fully the administrative activities of a hi-tech company,	f. investment group that specializes in the construction engineering sector.

7. While conducting a recent audit of a company's financial records, I became aware of	g. I spent much time in understanding market trends in administration management.
8. As a commissioned officer in our country's compulsory military service, I arranged the daily schedule of all the cadres in the platoon,	h. to the demand end — including customers' design-in status and demand forecasting.
9. Extensive laboratory training extended my ability to define specific situations,	i. think logically, collect related information, and analyze problems independently.
10. A large portion of my most recent work focused on increasing the company's revenue from the supply end — including the fab wafer start plan, IC assembly, testing, R&D design status -	j. served as a communication link between those cadres and higher-level officers, and coordinated missions within the unit, often assigned on short notice by those ranking officers.

2. Describing acquired knowledge, skills and/or leadership qualities / 描述個人所獲得的知識,技能及（或）領導特質，請參考以下範例：

範例 i.

Working in the quality control profession for nearly two decades has made me highly adaptive to change, responsive to sudden fluctuations in technological trends and flexible in acquiring diverse skills demanded in a competitive corporate climate.

範例 ii.

The hands-on experience in a laboratory setting provided me with numerous opportunities to corroborate what I had learned from textbooks and then extend that knowledge to an independent search for innovative solutions. I believe that laboratory research nurtured my skills that were previously lacking in experimental design, data evaluation and collaboration within a research group. These skills will now prove valuable in any future research project in which I am involved, hopefully at your company.

範例 iii.

Responsible for coordinating inter-departmental activities to effectively utilize the company's personnel and resources, I learned how to incorporate the seemingly polar approaches of different departments to solve a particular problem in a cohesive framework that drew upon the strengths of the departments involved.

範例 iv

By assuming such responsibilities, I acquired a breadth of experience in dealing with complex quality management-related topics. In addition to refining my ability to coordinate related activities, these experiences enabled me to examine an array of issues in economic, social, political and public financial contexts.

範例 v.

The occasional frustrations of slow progress in research strengthened my resolve to excel in the laboratory, making me more tenacious in spirit. Extensive laboratory work also exposed me to advanced experimental techniques and significantly improved my analytical skills and data collection capabilities.

第八個練習：

以下每句爲用來訓練有關管理師學術及專業訓練英文申請撰寫：描述個人所獲得的知識、技能及（或）領導特質。請把左邊及右邊的每一片語連結成一個完整句子。解答在本章最後。

1. Within our research group, I assumed	a. opened my mind towards emerging opportunities
2. The unique perspective I have acquired from conducting research in different countries and under varying circumstances	b. view of the intricate aspects of research and factors that spell success as an experimenter.
3. Considerable laboratory experience has given me a comprehensive	c. clarified my current educational needs and the career path that I should take to remain abreast of the latest trends in this increasingly hi tech market.
4. While finding the research process extremely satisfying,	d. resolve to excel in the laboratory, making me more tenacious.
5. Driven by the sense of achievement that follows a successful project, I have a personality	e. a leadership role by drawing others into discussion and exchanging opinions with our project leader.
6. Numerous work experiences have made my thinking more deliberate and.	f. responsibility and diligence when venturing into new projects.
7. My superiors have often commented on my adaptability, receptivity,	g. self-confident when encountering obstacles to the progress of research. I now view such obstacles as opportunities to adopt different approaches to the problem at hand.
8. Working in the finance sector for more than a decade has	h. has proven invaluable to my later professional endeavors.

9. Occasional frustrations with slow progress in research strengthened my	i. I also realize that continuous progress hinges on a persistent and tenacious attitude towards laboratory work.
10. Collaborating with others in several National Science Council-sponsored projects has made me more	j. that is conducive to thorough yet efficient laboratory work as part of a collaborative effort.

3. Samples of introducing research and professional experiences relevant to further academic or professional training

After receiving my Bachelor's degree in Information Management from Nanhua University in June 2003, I did not immediately start working, explaining my lack of professional experience. However, I strengthened my computer skills during undergraduate school. During my university studies, I assisted my classmates and relatives with computer related problems. I not only provided assistance with computer related problem, but also learned new techniques from doing such work. When I encountered difficulties, I had to study and solve them. I enjoyed a brief period of practical training as a sales representative at Acer, as well as practical training at National Chung Cheng University after graduating from university. University studies also exposed me to many software programs. I also learned a great deal about other programs and operating systems from my practical training. These programs helped me to solve many questions and work more quickly. Knowledge must be obtained from libraries or other sources, which takes considerable time and effort; the veracity of the data thus obtained is not even certain. An increasing number of data methods and tools are available from the Internet, including interview data. Such data are ubiquitous, and can be found with minimal time and effort, enabling us to read this data quickly and efficaciously. Many programs have been developed, helping researchers to perform counting and analysis. These data and programs improve research reliability. Therefore, I obtained a great deal from study and practical training, helping me considerably and supporting me in my research work. An increasing amount of data and programs are available on the Internet, which will help future academic research. I will continue to keep myself up to date through such study.

Following university graduation, I joined the army as a commissioned officer. I was initially responsible for handling administrative work involved monthly in the physical examinations of six thousand new soldiers out of a large squadron of around ten thousand military personnel. In this capacity, I was directly responsible to a general. The environment provided me with many opportunities not only to coordinate with experts in diverse fields, but also to familiarize myself with how a large-scale organization operates. These valuable experiences enabled me to comprehend how slogans, mission statements and corporate strategies can be transformed into action plans. My second job responsibility was managing the activities of nearly 160 soldiers in a platoon. Blending my previous experiences with the task at hand, I not only realized how students lack experience in adopting different models and operations in medium-sized and large-scale organizations, but also comprehended why basic level employees always become discouraged with many management methods that are neither applicable in actual business practices nor innovative. Administrators in these organizations have various management styles and ways of implementing tasks. For instance, large-scale organizations tend to emphasize legal procedures, but lack flexibility and efficiency. Whereas medium-sized organizations focus on leader's personality and style, they lack sufficient resources and support. The above experiences have allowed me to acquire management core concepts and strengthen my communication skills. After retiring from the professional military service of nearly four years, I served as a research assistant in the Society of Emergency Medicine, R.O.C., which allowed me not only to see the connection between theoretical knowledge and actual business practices, but also to realize how results from academic investigations can influence governmental medical policies.

Having devoted myself to developing information systems in the semiconductor industry for over a decade, I developed a particular interest in enhancing work productivity via use of the latest information technologies. I also spent considerable time in researching system integration for manufacturing applications on UNIX-based systems. Critical thinking skills developed during my undergraduate and graduate training enabled me not only to explore beyond the initial appearances of manufacturing-related issues and delve into their underlying implications, but also to conceptualize problems in different ways. Notably, I initiated an intranet project as a section manager at MOSEL Corporation. This responsibility oriented me on how to lead a team and thoroughly understand various software development processes. While participating in several MOSEL group projects, I learned how to address supply chain-related issues in order to broaden my perspective on potential applications of finance and decision making; these areas are now my main focus of interest. I am confident that my working experience in software development will equip me with the necessary competence to effectively address problems in the workplace. Moreover, my project management experience has enabled me to carefully deal with others and resolve disputes efficiently. My love of challenges will enable me to satisfy constantly fluctuating customer requirements in information integration projects.

E. Describing extracurricular activities relevant to academic or professional training／描述與學術及專業訓練有關的課外活動

1. Introducing an extracurricular activity that one has participated／曾參加的相關課外活動介紹，請參考以下範例：

範例 i.

As a volunteer for the Society of Quality Management during graduate school, I was responsible for organizing academic workshops, managing departmental publications, performing various administrative tasks, and organizing book exhibitions. This experience instilled in me the importance of synergy and efficiency when collaborating with others.

範例 ii.

During my campus life, I spent time developing my own interests and focusing on personal growth beyond academia.

範例 iii.

In addition to acquiring professional knowledge at graduate school and work, I cultivated my social skills and a collaborative spirit through mountain climbing in a campus-based organization. Climbing allows me to release pressure from work by pushing my physical limits and improving my health.

範例 iv.

As an undergraduate, I organized a private consulting group of university students who tutored junior high school students in an array of subjects. Despite initial difficulties in establishing this group, I was able to recruit a large number of tutors, rent a location, purchase teaching materials, and design a daily study schedule so that the students could keep up with the curriculum. The community response was overwhelming, as evidenced by the large number of students who enrolled in our tutorial courses.

範例 v.

An outstanding manager must balance academic and social skills. Therefore, I actively participated in our university's student union, the most reputable student organization on campus, to apply systematically my knowledge skills and broaden my horizons. After several years of participation, I served as a vice-president of the association, in which role I was responsible for various tasks such as coordinating student welfare and communicating students' needs to pertinent school authorities, holding campus events, and contacting student unions of other universities to set up collaborative events.

2. Highlighting the acquired knowledge, skills or leadership qualities that are relevant to academic study or professional training／強調和求職相關的已有知識技能或領導特質

範例 i.

While leading the campus orchestral group, I learned how to motivate group members to hone their individual talents for a common goal. By dealing with the occasional frustrations of fiercely individualistic orchestral members when trying to reach a consensus has made me a strong leader, I was able to listen patiently to individual concerns and trying to incorporate them in a group consensus. I believe that that role was good training for what I will encounter in the workplace.

範例 ii.

Participating in the university Heart club, a social service organization, helped me to nurture leadership and collaborative skills as an organizer and coordinator of many activities. ALE, an acronym for "Actual Living English", not only strengthened my listening, speaking, reading and writing skills, but also allowed me to make new friends from different cultures and express my feelings in another language, which is challenging for me. Experiences learned by participating in extracurricular activities have greatly enriched my life by enabling me to achieve the necessary balance of academic and social skills.

範例 iii.

The numerous extracurricular activities in which I participated in at university, taught me how to organize activities competently, express my opinions clearly, communicate with others effectively, transform conflicts into constructive situations that promote collaboration, and cope with failure by analyzing the underlying problems that prevent success.

範例 iv.

Extensive global travel for business and recreation has made me more receptive to occasionally opposing values of different countries. Closely observing and comparing different societies, rather than making sweeping generalizations about a particular country or group of people, have allowed me to observe various management practices in different cultures.

範例 v.

As a strong academic performer, I acquired solid communication and coordination skills while leading our department's student association for two years. The collaborative skills demanded by the position speak to another trait of my personality and, in doing so, won me the respect of my peers, faculty members and staff.

第九個練習：

以下每句爲用來訓練有關管理師學術及專業訓練英文申請撰寫：強調和求職相關的已有知識技能或領導特質。請把左邊及右邊的每一片語連結成一個完整句子。解答在本章最後。

1. Acting as a liason between the student body and the local city government not only furthered my organizational abilities and ability to handle a crisis, but	a. life by enabling me to achieve the necessary balance of academic and social skills.
2. Dealing with the occasional frustrations of fiercely individualistic group members when trying to reach a consensus has made me a strong leader.	b. allows me to mix well with others from diverse backgrounds.
3. Experiences gained through my participation in extracurricular activities have greatly enriched my	c. also made me a perseverant and punctual individual who can cope with unforeseeable circumstances.
4. Volunteer social work allowed me to view others' circumstances objectively,	d. I was able to listen patiently to individual concerns and tried to incorporate them in a group consensus.
5. Coordinating club activities and closely collaborating	e. but also instilled in me the necessity of systematically and methodically making concerted efforts with others , something which is vital in a research career.
6. My congenial personality, nurtured by these extracurricular activities,	f. excellent communicative skills and an assiduous working attitude to implement club activities successfully.
7. These experiences strengthened my communicative, organizational,	g. leadership, and management skills.

8. This challenging work made me confident in my ability to achieve goals	h. often difficult for someone who has been engrossed with academia since childhood.
9. These activities instilled in me the importance of developing an outgoing personality,	i. by systematically applying knowledge skills.
10. Working part-time as a network manager at an Internet service provider not only taught me a great deal about Internet communication,	j. with other volunteers provided a valuable point of reference when I later organized my Master's thesis.

◎Samples of describing extracurricular activities relevant to academic or professional training

Participating in many extracurricular activities during university enabled me to acquire many practical skills and form many friendships. Such experiences not only made my life more intriguing, but also exposed me to many aspects of life that I could not learn in a classroom. While learning how to get along with others and strengthen my conversational skills, I obtained many communication skills through the many activities I joined. Through the many friendships subsequently formed, these companions made me happy and taught me how to resolve difficulties collaboratively. For instance, while studying, my friends often encouraged me to excel by working with me on class assignments and preparing for examinations. As for my interests, I enjoy listening to music, playing basketball and reading, which not only helps me pass time but also benefits me personally. Playing basketball not only strengthens my physical stamina, but also makes me more perseverant and determined - skills which are especially helpful in academics. Perseverance allows me to resolve classroom difficulties. For instance, strengthening my reading skills not only through textbooks, but also through novels, magazines and newspapers keeps me abreast of societal trends and increase my knowledge skills. Such knowledge is not trivial, possibly helping me to resolve some future problem that textbooks provide no answer. This reading pastime also expands my thinking, ultimately making me more mature. Likewise, listening to music calms me, a form of medication for my soul. Without music, my life would have a deep void. Although the above extracurricular activities took much time and energy, I gained many intangible treasures that complemented my studies.

I enjoy reading, playing basketball and watching movies in my leisure time, not to mention venturing outdoors, listening to music, camping and traveling. Of these pastimes, I especially love traveling. In my leisure time, I always journey with my junior high school classmates to the countryside to enjoy nature. I also organized an overseas trip for university classmates. Working part time and studying simultaneously explain why I found employment during the daytime and pursued a bachelor's degree at night. My knowledge of finance and accounting comes from my experience in bookkeeping, analyzing financial statements and controlling cash flow…all of which come from working in a finance department for nearly three years. Employment in the financial department and a strong desire to become a professional analyst in this field have prepared me for graduate study in the Institute of Business Management at Yuanpei University. Above experiences have not only made my life more meaningful, but also instilled in me the importance of the personal connections that will benefit my career. The many extracurricular activities that I participated in during university gave me the necessary balance of developing academic and social skills simultaneously. I often served as a representative or spokesperson in class organizations, subsequently making many new friends and building my self-esteem. In addition to a solid academic background, a good manager should have strong communicative, organizational and management skills. While increasing my interest in this field, I have become confident of my ability to achieve career goals. Further strengthening my solid knowledge skills will enable me to remain abreast of the latest trends in finance-related topics.

As a university freshman, I played on the campus intramural badminton team, which was both relaxing and challenging. I was also a member of the archery club as a sophomore. Despite no previous experience, I got caught up in the energy and strong sense of camaraderie. Besides exercising in the gymnasium two and three times weekly, I also enjoy reading and watching movies, of which, The Lord of The Rings and Harry Potter are my favorites. I especially enjoy reading books that are necessarily related to studies. During the summer break of my sophomore year, I worked part-time at SOGO Department Store. During this period, I learned much about customer relationship management and how quality service and excellent product quality can influence the purchasing decisions of customers. Additionally, product promotion is an important reflection of corporate achievements. Moreover, organizations must remain attune of changes in consumer lifestyle trends owing to the large potential profits. In summer, working in the department store taught me the importance of close collaboration among colleagues. During the summer break of my junior year, I served in a practicum internship at Lo-dong Pohai Hospital in Yilan, Taiwan. The internship allowed me to work in various departments, including materials requisition, human resources, patient records and documentation, as well as public relations. Learning the daily routine of hospital operations has provided a valuable reference for future employment. To supplement my studies, I traveled to Taipei twice weekly to attend a statistics course, which eventually enabled me to pass the graduate school entrance examination.

During university, I participated in several extracurricular activities. My teachers often encouraged me to take part in the competitions, such as composition writing or Chinese calligraphy, which I often did quite well. Such competitions strengthened my stamina and perseverance. More than just receiving commendation from teachers and recognition, I view these events as quite memorable. Additionally, a classmate who had been elected to a position in the student association put me in charge of taking meeting minutes and publicizing activities among the student body. Belonging to a team of classmates enabled me to form many friendships and build my self-confidence. Besides a solid academic background, I should foster my communicative, management and organizational skills. I especially need to learn how to effectively handle crisis situations. Such practical experiences can not be learned from sitting in a classroom. I also enrolled in an extracurricular activity on etiquette in commercial settings; I served as a teaching assistant as well. Moreover, during our university's anniversary celebrations, I served on the reception committee to receive honored guests. In sum, undergraduate study provided me with many opportunities to lead other classmates. Given the satisfaction of my teachers with my progress, I was assigned to several leadership positions, enabling me to put into practice what I learned in class. I firmly believe that proper etiquette is essential for smoothly implementing a task. While increasing my interest in the field of communications, above experiences made my confident of my leadership skills.

As time progresses, university students face increasing pressures in daily life: academic, economic, social and sexual ones. Students often struggle in coping with such pressures, tragically resulting in suicide for some or failing either academically or in personal relations for others. While suicide has received much media attention, its incidence rate is actually on the decline. Therefore, for university studies, extracurricular campus activities are essential to releasing some of the above pressures. For me, sports are my passion, especially badminton and golf. In these extracurricular activities, I learned how not only to refine my badminton and golf skills, but also to release daily pressures. For me, releasing daily pressures means forgetting things that upset me, clearing my mind of distractions and maintaining my physical stamina. After studying diligently for a certain period of time or experiencing a personal setback, I will ask my classmates to join me in a badminton match in the university gymnasium. Sports are especially helpful in raising one's spirits. Besides participating in extracurricular campus activities, I often go swimming during university breaks, take my dog for a walk, watch a movie at the cinema and read literature not related to my studies. In sum, although capable of keeping up with my academic studies, I equally need natural ways of relieving pressures from daily life. While societal pressures demand more of modern individuals, students must balance their academic studies with healthy releases of pressure.

【解答】

第一個練習：1.C 2.J 3.F 4.A 5.H 6.B 7.G 8.E 9.D 10.I
第二個練習：1.C 2.F 3.H 4.A 5.I 6.J 7.B 8.E 9.G 10.D
第三個練習：1.E 2.C 3.G 4.B 5.F 6.A 7.D 8.H 9.J 10.I
第四個練習：1.G 2.I 3.A 4.J 5.H 6.B 7.C 8.F 9.D 10.E
第五個練習：1.E 2.H 3.B 4.G 5.A 6.C 7.D 8.F 9.J 10.I
第六個練習：1.F 2.C 3.E 4.A 5.G 6.D 7.B 8.J 9.H 10.I
第七個練習：1.D 2.G 3.A 4.F 5.B 6.C 7.E 8.J 9.I 10.H
第八個練習：1.E 2.H 3.B 4.I 5.J 6.A 7.F 8.C 9.D 10.G
第九個練習：1.C 2.D 3.A 4.H 5.J 6.B 7.G 8.I 9.F 10.E

【參考文獻】

柯泰德（2002）。《有效撰寫讀書計畫》。新竹：清蔚科技公司。

Appendix

參考範例 i.

Study Plan

Despite working in the information sector, I majored in Electrical Engineering at university. Nevertheless, I am fascinated with business management-related topics. To more fully understand how information technologies can enhance business management, I am pursuing a graduate degree in Business Administration at Yuan Pei University (YPU).

I have worked on developing information systems for nearly a decade, thus becoming quite familiar with computer technology and semiconductor processes. The strong research fundamentals acquired from my work experience have prepared me for advanced study on how information systems can assist managers in making the optimal decision.

My graduate research will hopefully center on how to perform mathematical calculations, such as use of gray and fuzzy methods, to simulate business decisions more effectively. Furthermore, I plan to develop an expert information system.

As for expert information systems, I have studied much on lot dispatching, entity feedback systems and cost accounting information systems. I am particularly interested in how to make an accurate, effective and efficient database system. Thus, I wish to further my knowledge not only on business management-related topics, but also on mathematical calculations for use in real time control, marketing and decision making.

Although not renowned for its Business Administration program, YPU recently established a graduate school in this discipline. Faculty in the graduate school has published widely in this field, such as the university president, whom is recognized for his research in gray theory and the marketing professor, whom has a solid background in this field. The technical writing instructor is also reputed in his field. This explains why I chose YUST for graduate study. My solid background in information technology has prepared me to meet the rigorous challenges of Yuanpei's challenging graduate programs. Graduate study will undoubtedly equip me for success as a business manager someday.

After completing our country's compulsory military service of nearly two years, I started working in an information-integrated company to start my career. The company's comprehensive operations broadened my horizons as to the dynamics of working in the hi-tech sector. I first learned how to maintain hardware, construct a network and, finally, learn the principles of how an information system is manufactured. Moreover, I participated in several MIS projects on ensuring the software quality and bringing a new system to the production line. In addition to nurturing my problem-solving and basic knowledge skills in software engineering, this work experience exposed me to the complexity of interacting with individuals in an intensely competitive environment. I later served in the MIS Department of a semiconductor manufacturer, coming into contact with the latest information systems at that time. After striving diligently to familiarize myself with the latest information technologies, e.g., programming skills, database management and UNIX environment, I was later promoted to section manager in 1999. While working in the semiconductor company, I not only kept the MIS system running smoothly without any crashes, but also expanded the system capacity to satisfy user's requests and reduce overhead costs. Based on principles of efficient development, easy maintenance, stable operations and low operational costs, I spent half a year in developing a intranet system for the company. The software development of this intranet exposed me to both objective-oriented analysis

(OOA) and objective-oriented programming (OOP); I also became proficient in adopting related methods. After leaving the semiconductor industry, I began working for an information-integrated company again. As a project manager, I am responsible for performing systems analysis and negotiating with clients. More than integrating information technologies in the company through software development, I must refine my interpersonal skills so that I can more effectively lead a development team and achieve on-time delivery of our company's products and services.

Taiwan has become a globally leading manufacturing center, especially in computer related products. Although information systems imported from abroad are extensively adopted in manufacturing, the analyzed information is insufficient to assist staff in completing their tasks effectively. Thus, working overtime is still critical for making a high quantity of products rapidly. This explains why death incurred from overworking has been often discussed recently. Growing up in such a competitive environment, we should strive to work efficiently. Thus, how to upgrade those information systems in order to correspond to Taiwan's unique manufacturing process and management culture is increasingly important. More than information technology, an excellent system must also consider management issues. Therefore, I decide to purse a graduate degree in Business Management in order to become an expert in developing information systems.

Having devoted myself to developing computer information systems for over a decade, I am well aware that no information system can be implemented problem free. Restated, the end user must always spend much time in modifying some functions that correspond to practical operations. Even after the system has gone online, the system must be continually enhanced or modified owing to company re-engineering, re-organization, ergonomic issues and consumer demand. Thus, I would like to equip myself with stronger analytical skills to accurately identify what a system requires to develop, enhance or modify an information system effectively.

Although my work experience has focused on collecting data and overseeing workflow control on a production line, my graduate school research will hopefully allow me to develop a database method that integrates data mining and knowledge management approaches. I would thus like to study how machine learning, statistics and visualization technologies can be adopted to identify and present knowledge that humans can easily comprehend. I will also effectively use software development technologies to ensure that an information system performs optimally. I firmly believe that combining academic concepts with my knowledge expertise from previous employment will enable me to rise to the arduous challenges of performing graduate level research.

I always remain abreast of the latest governmental policies and economic trends by reading several newspapers and magazines regularly. I also discussed with my friends on their career developments, including helpful ideas or technologies, especially on how to enhance existing information systems. Besides, I enjoy reading Chinese ancient literature, such as The Art of War by Sun Tzu and Tao Teh Jing by Lao Tzu, authors whom are distinguished for Chinese philosophy worldwide and often inspire me in tackling management issues. In addition to reading many of their writings, I often search for information that is not available on Internet search engines. In addition to deriving much pleasure from reading, I have participated in a martial arts school. This discipline not only regulates my health, but also cultivates my inner spirit.

While certain individuals may view me as overanxious to assimilate much information, I constantly distinguish between academic/work reading and pleasure reading. I also enjoy sharing what I have learned with my friends. Sharing with others can supplement my knowledge and cultivate interpersonal and communication skills. In the era of a seemingly overload of information, I believe that effective collaboration is essential to success, explaining why I am always fascinated with management issues and pursuing a graduate school degree in Business Management.

參考範例 ii

Study Plan

Disputes over medical treatment have intrigued me not only because they are interesting, but also because physicians and patients experience similar medical treatment problems. I am especially fascinated with diagnosing and treating medical problems owing to their interesting nature. Additionally, many events concerning medical treatment have occurred in recent years, especially how to use and inject a curative. Treating medical problems is a major concern of daily life While many events have occurred in Taiwan related to treating medical problems, local hospitals still lack knowledge on how to distinguish between different diseases or maladies. Also, patients have many questions on how to use curatives properly. Proper instruction on how to use a curative from a physician or nurse would help prevent unnecessary hazards. In addition, physicians and nurses must have specialized knowledge skills to carefully diagnose a patient' s illness. Most domestic drugs have English language instruction for proper usage. However, if the drug contained instruction in Chinese for its proper use, then the likelihood of filling a faulty prescription would decrease, thus ensuring individual safety in dealng with curatives.

I received a bachelor' s degree from the Healthcare Management Department at Yuanpei University (YPU) in 2003. The departmental curricula provided me with a theoretical and practical understanding of management courses, especially on how to accumulate statistics and effectively manage sales and financial affairs While gathering statistics and effectively managing financial affairs proved to be the courses during university, I realized that they were not especially difficult after diligently studying the topic at hand if I only remained attentive or diligently reviewed different briefs. I am especially interested in researching business process reengineering, which is interesting for me because I can understand how a business operates or how a hospital may lose its competitiveness. How to effectively solve problems involving a hospital's competitiveness is particularly intriguing. I am intrigued with this aspect of quality management. After completing my bachelor's degree at YPU, I secured employment at the university health care center. The center is currently involved in a campus wide anti-smoking campaign. I participated in an anti-smoking orientation program in Taipei on how to combat smoking among students. This program brought me into contact with anti-smoking consultants and oriented me on how to effectively market this campaign among students. My graduate school research will hopefully center on combining theoretical and practical approaches to adopt marketing practices when combating tobacco use.

University also provided me with the opportunity to work in a practicum internship at Chimei Hospital during summer vacation, where I assisted in personnel matters and administrative tasks. This year, I served in a practicum at a Hsinchu-area hospital in which I participated in a hospital evaluation . This work allowed me to understand how a hospital evaluates its personnel and current policies. I also worked in a restaurant as well. Up until graduating from university, I also worked in the campus health care enter. This interesting job focuses on ways in which to combat smoking among students and how to properly orient the student body on its dangers. The target group also includes primary to high school students in Hsinchu City. The Health Care Center heavily prioritizes creating a nonsmoking environment on campus.

As for my extracurricular activities, I participated in campus wide anti-smoking activities sponsored by the R.O.C. Bureau of Health Promotion Department. In addition to offering general orientation on the dangers of smoking, the activities also informed students on tobacco manufacturers' deceptive motives and marketing approaches. These activities profoundly impacted me to the point that my graduation thesis addressed how marketing is used in this area. As for my personality, I am easy to get along with. During university, although my academic marks were undistinguishable, I always strived to study diligently despite my personal limitations. When working on a specific task, I always act quickly to efficiently resolve problems. I am also able to function well under pressure, which makes me a direct individual who does not evade pertinent questions with confidence. Above qualities make me receptive to learn new concepts, which will prove to be valuable assets when conducting graduate school research and completing my master's thesis.

參考範例 iii.

Study Plan

Hopefully, my graduate school research will encompass four disciplines: physiology, psychology, organizational behavior and medical supply equipment. However, healthcare management professionals often have difficulty in thoroughly understanding how each unit is related. Healthcare management professionals must determine how to offer not only a humane environment for medical patients, but also a long-term strategy that emphasizes a holistic approach to quality care throughout one's entire life. To address this situation, non-profit organizations hold the majority market share of the long-term care segment, followed by governmental institutions and then hospitals. Inflexibility, inequity and a lack of supervision often occur in the long-term care segment despite the increasing number of organizations that provide services. However, they lack effective management in confidentiality laws and uniformity across the various governmental authorities in charge of related affairs. Many families without adequate manpower have their elderly relatives remaining in hospitals for extended periods, resulting in a tremendous waste of national health resources and great economic pressure on families. In other circumstances, inhumane treatment and inadequate living conditions have occurred in the long-term care segment by organizations that operate below national standards. Moreover, losing one's physical functions without an appropriate means of venting one's pressure has resulted in an increasing number of elderly without a social position and prone to psychological ailments, a low living standard and eventually many societal problems. What measures can be adopted to ensure that appropriate resources are devoted to ensure that the above problems can be solved? From a microcosmic perspective, administrative management aims to provide high quality and relatively inexpensive products and services by developing and implementing models and standards that incorporate the above four aspects of this topic. I am especially interested in this research topic.

I received a Bachelor's degree in Healthcare Administration from Chungtai Institute of Health Sciences and Technology in 1998. The departmental curricula included hospital administrative courses, as well as a solid understanding of professional knowledge within the medical sector, especially internal affairs and public relations, as well as external medical policies and customer relations management (CRM). While working in a practicum internship for a medical supply company, I found an actual working environment to be much more complex than what I was taught in the classroom. After retiring from professional military service of nearly four years, I decided to pursue business-related issues by further attending not only several introductory courses, but also many conferences that addressed topics in this field. Moreover, I served as a research assistant in the Society of Emergency Medicine, R.O.C., which allowed me not only to see the connection between theoretical knowledge and actual commercial practices, but also to realize how results from academic investigations can influence governmental medical policies.

Following university graduation, I joined the army as a professional soldier and officer. I was initially responsible for handling administrative work involved monthly in the physical examinations of six thousand new soldiers out of a large squadron of around ten thousand military personnel. In this capacity, I was directly responsible to a general. The environment provided me with many opportunities not only to come into contact with hospital staff and physicians, but also to familiarize myself with large-scale operations. These valuable experiences enabled me to comprehend not only why confusion often occurred in a health environment, but also why health care administration students become discouraged with the profession.

My second job involved managing the activities of nearly 160 soldiers. Blending my previous experiences with the task at hand, I realized how students lack experience in adopting different models and operations in medium-sized and large-scale organizations. Administrators in these organizations have various management styles and ways of implementing tasks. For instance, large-scale organizations tend to emphasize legal procedures, but lack flexibility and efficiency. Whereas medium-sized organizations focus on a leader's personality and style, they lack sufficient resources and support. The above experiences have allowed me to understand fundamental concepts of becoming a professional in the health care and management sector.

During university, campus life was filled with participation in several student associations. For instance, I was actively involved in the student activity center (SAC) for three years. SAC was largely responsible for coordinating the administrative activities of sixty-five student associations, such as distributing funds, supervising and supporting activities, providing consultation and sponsoring large-scale activities. At SAC, I was in charge of furnishing and operating stage lighting and acoustics for campus events. This position offered ample opportunities for me not only to come into contact with people with diverse backgrounds and various activities, but also to familiarize myself with writing proposals and implementing a wide array of skills. By adopting many of my experiences from serving in the Society of Emergency Medicine, R.O.C., I became aware of how important writing skills and related administrative skills are because SAC's approach to planning a framework and implementing operations closely resembles that of academic research. During university, I also served as a student association leader, in which I cooperated with the government sector to hold small-scale activities. This position not only furthered my organizational abilities and ability to handle a crisis, but also made me a perseverant and punctual individual who could cope with unforeseeable circumstances.

Whether at work or school, my colleagues and friends have often commented on my diligent and trustworthy character. When assigned a task, I carefully arrange a flexible schedule that can be adjusted in case a crisis or need to change arises. Work experiences and extracurricular activities with the student association have definitely fostered my communicative skills and ability to achieve accuracy and efficiency. For instance, holding conference meetings for collaborative efforts is an essential communication form in daily work despite its occasional tedious and inefficiency nature. My ability to effectively use time in conference meetings reflects my strong desire to identify the interests of individual collaborators while actively pursuing a common goal that will strengthen our knowledge skills and expertise. In particular, I have come to understand the merits and limitations of various methods in a business enterprise to attain a clearly defined objective. In addition to a solid academic background, a good manager should have strong communicative, organizational and management skills. I am confident that I possess these qualities.

Unit Eight

Delivering Effective Oral Presentations

有效進行管理英文口語簡報

簡 介

本單元主要介紹如何有效進行管理專業類的口語簡報，包括簡介、本體及結論部分的有用詞彙。本單元也同時介紹不同型式的口語簡報，包括：預測市場趨勢、產品或服務研發、專案描述、公司或組織介紹、組或部門介紹、工業介紹。

一、口語式演講報告的結構要素

A. 開場

1.問候致意和演講主題介紹

· Good afternoon, ladies and gentleman. I would first like to thank the organizational committee for inviting me to be here with you today.
各位先生、女士們，下午好。我首先要感謝機構的委員會，能在今天邀請我來這裡和大家見面。

· I am delighted to have the opportunity to be with you today and provide a general overview of...
我很高興在今天能有這個機會，在這裡爲大家提供一個概念，是關於……

· I would like to thank you for giving us the opportunity to visit your company and discuss with you...
我必須要感謝各位，願意提供我們這個機會來這裡拜訪大家，以及共同探討……

· I am pleased to have the opportunity to give a presentation on...
我很希望能有這個機會來爲大家做一個說明會在……

· I'm really pleased to be here and give you a glimpse into...
我非常希望能在這裡提供各位一個簡單的概念來進入……

· I'm going to talk about...
我將要討論的是關於……

· The objectives of today's presentation are to provide you with a basic understanding of...
今天這個演講的目的，主要是爲了提供各位一個關於（某主題）的基本認識……

2.演講報告的大綱摘要

- I will give this presentation in five parts.
  我把今天的演講內容分成五個部分

- I'll divide today's talk into four parts.
  今天討論的內容，我將會用四個部分帶入

- I'd like to cover five areas today.
  我希望今天能探討到五個重點

- Today's presentation will be given in five parts.
  今天的演講報告將會被切割成五個部分

- Before getting into the main subject, I would like to briefly introduce...
  在討論到今天的主要題目前，我希望能大概的簡約介紹一下……

- I'll start out by briefly introducing...
  我會先用簡約的介紹方式來爲這個演講報告起個頭……

- First of all, I will provide you with a general introduction of...
  接下來，我將會把重點放在……

- Next, I would like to focus on...
  緊接著，我要概略性的談到……

- Moving on, I'd like to briefly touch on...
  接下來，我將會討論爲什麼……

- Next, I will discuss why...
  接下來，我將會詳盡的解釋關於……

- Next, I'll elaborate on...
  接下來，我將把重點放在……

・Next, we will look at...
然後，我將會帶到……

・Then, I will move on to...
接下來，我將會討論到關於（某重點）的特性

・Next, I will discuss the nature of...
這個演講報告的第三個部分攸關於……

・The third part of this presentation deals with...
第三個部分攸關於……

・And finally, I would like to make some concluding remarks.
我要在最後指出，某些部分在未來的可能性和挑戰

・I'd like to finally point out some of the future challenges of...
我希望能指引出某些未來的挑戰

・Finally, I would like to point out some of the future directions...
最後，我希望能指引出某些未來的方向

B.組織架構

1.第一個論點

・Before I get into the main subject, would you allow me to briefly introduce…?
在我們還沒有進入討論的重點以前，可否允許我先大致的簡單介紹一下……？

・I would first like to give you an idea of...
我想先給你一個關於（什麼）的概念

・First of all, why should we...?
在此之前，爲什麼我們需要……？

・Let's begin with...
讓我們開始（某主題）

・I'd like to begin by making some introductory remarks about...
我希望能先以重點式的介紹，再來開始討論這個主題

・It would be best to start off by making some general comments about...
如果，以一般社會大眾的意見，來開始這個主題，我想是最好不過的了

2.第二個論點

・The next consideration is...
接下來必須考慮到的是……

・Now allow me to...
現在請容許我……

・Let's now take a closer look at...
就讓我們現在一起深入……

・Now let's turn to some of the specific areas that...
現在讓我們開始轉入一些更加明細的主題範圍

・This now brings us up to the second part of today's presentation...
這將會帶領我們大家一起進入，今天這個演講報告主題的第二個部分……

・This brings us up to our second point:
這將會帶我們進入第二個論點

3.附加的論點

・Shifting to the next part of today's presentation, I'd like to briefly touch on...
接下來，將轉移到今天這個演講報告的重點，我將概要性的帶到……

・Now let's turn to...
現在讓我們轉到……

・Let's now move on to...
讓我們現在一起看看……

・I'd now like to shift to...
我現在想要將話題轉移到……

・I would like to provide you with some extra information about...
我想要提供你一些關於（某主題）額外的資訊……

・The next point that I would like to bring up is...
接下來我想要把重點帶到……

・Next, let's briefly touch on...
跟著，就簡單的談談……

・Now, I'd like to ask the question: ...?
現在，我想要問的問題是……

・Moving on, I'd next like to highlight...
跟著下來，我想要強調以下的……

・Of course, all of this has special significance when applied to my third point:
當然，這些所有的重點都有它們的意義，尤其是套用在我的第三個論點時

4.內建式的摘要

- At this point, I'd like to sum up the results of...
  在這個論點，我想要將所有的結果做個總合

- Before going further, I'd like to summarize what we have already covered.
  在繼續下一個部分之前，我想要把所有之前的討論做個概括總結

5.最後的論點

- Before closing, I would like to sum up...
  在結束之前，我想要總合一下……

- Before I close, I must say a few words about...
  在我結束之前，我必須先說明一下，關於……

- Before concluding today's presentation, I would like to briefly mention...
  在爲今天的演講報告做出總結論之前，我想簡單的提一下……

C.結論

1.將所有重點做出概要

- In concluding my presentation today, I would like to emphasize the following.
  在爲我今天的演講報告做出結論前，我希望能強調一下關於以下

- From the developments reported above, I would like to make some concluding remarks for today's presentation, which will hopefully shed some light on...
  從今天所有談到的報告內容中，我想要把重點做出些結論來指出些許方向

- Finally, I would like to make some closing remarks for today's presentation.
  最後，我希望能爲今天的演講報告做一個結束前的總合整理

- In concluding my presentation today, I would like to point out...
  在結束我今天的演講報告和做出結論之前，我想要指出……

- I'd like to close today's presentation by pointing out some of the future challenges that we are faced with...
  我希望以重點式的整理出我們將來會面對到的可能性的挑戰，來爲今天的演講報告做個結尾……

- Finally, I would like to sum up by pointing out some of the future directions of...
  最後，我希望能在結束前，以指出未來方向的方式做個總結

2.感謝各位觀衆

- Thank you for your kind attention.
  謝謝你們給予誠意的專注力

- Thank you for coming here today.
  謝謝你們今天的到場

- Thank you for your patience.
  謝謝你們耐心的聆聽

- Thank you for your time today.
  謝謝你們今天撥出時間來參與

- Thanks for the opportunity you have given me to be here with you today.
  謝謝你們給我這個機會，讓我今天可以在這裡和大家見面

- Ladies and gentlemen, it has been a great pleasure for me to be here today.
  女士和先生們，能在這裡和各位共度是我至高無上的榮譽

3.邀請在場人士發問

・I see that we have a few minutes remaining on the program. I welcome your comments or suggestions.
我看了看，我們現在還剩下幾分鐘，我非常歡迎在場的各位能提供你們的觀感和提供出你們寶貴的意見。

・If you have any questions, please feel free to ask.
如果您有任何問題，請不要感覺拘束，請自由發問。

・I' d now be happy to answer any of your questions.
不論任何問題，我都很高興爲大家回答。

二、口語簡報的形式

A.預測市場趨勢，開發此產品的原因：

1.科技層面：(1)相關產品；(2)潛在利潤

・A growing elderly population will increase the rate that individuals retire, with a shrinking younger generation to support the elderly. Therefore, understanding how the quality of life of the elderly and the retail housing market for this age group are related has received increasing attention given current trends in Taiwan' s aging society.

・Living standards in Taiwan have dramatically increased in recent years, as evidenced by average per capita incomes surpassing $US 10,000 and the island' s position among five of the largest holders of foreign exchange reserves worldwide.

2.財務層面：開發領域的市場趨勢

- While Taiwan's elderly individuals over 65 years old in 2003 comprised 9.2% of the entire population, this figure will increase to 15.9% by 2021. To effectively respond to this population shift, policy makers and various health care professionals are concerned with how to effectively coordinate available resources.

- Income generated from insurance premiums in Taiwan has maintained stable growth in recent years, surpassing the NT$100,000,000,000 mark in 2002; of which, 51.23% came from commercial banks. This figure reflects the continuous development of the bank-sponsored life insurance segment.

3. 產品開發的組織策略和方案：(1)策略的界定（包括最高目標及主要重點）；(2)執行策略方案的界定

- Effectively promoting the use of solar energy requires an emphasis not only on its continuous development to gain public confidence in its ability to provide for energy consumption in the future, but also on its applicability in a diverse array of product appliances and facilities.

- To foster its competitiveness, the domestic electronic communications sector must strengthen its research capabilities, develop lower electromagnetic wave technologies and attract technology professionals with expertise in multidisciplinary fields.

4.市場調查：(1)產品的商業潛力；(2)產品運用的範疇

- According to Taiwan's Annual Real Estate Report, an increasing number of individuals will live independently in rented apartment units located in housing complexes where the elderly live exclusively. While such a trend is aimed at upper income elderly, marketing efforts should be made to promote this future living trend among this growth sector in Taiwan and gain societal acceptance in a traditional society where the younger generation is normally expected to care for the elderly.

- According to a recent market survey, although only around 20% of all disabled elderly in Taiwan receive institutional-based care, the market demand for institutional-based care is 30% (Department of Health, 1997). The 10% difference is equivalent to a market scale of at least 18,000 individuals. Moreover, the annual growth rate for disabled elderly in Taiwan is nearly 20%.

範例 i.

The widespread availability of advanced technology products has diversified the rapidly expanding consumer market in Taiwan. More than an improved design of an older brand, new products reflect how consumer preferences have dramatically changed. The digital camera market is no exception, as evidenced by its concern for product specifications and customer satisfaction level. As the market scale enlarges, breakthroughs in digital camera technology appear constantly. Sony is an especially popular product in Taiwan given its strong brand recognition and high prestige. The brand image remains strong owing to Sony's ability to increase customer satisfaction levels by ensuring that product design complies with consumer recommendations. Representative of the company's attempt to encourage individuality, the Sony Cyber-shot digital camera offers unique functions that allow users to tailor photographs according to their preferences. Despite the fierce competition among digital camera manufacturers, Sony continues to attract customers by maintaining a specified price elasticity and continuously upgrading the functionality of its digital cameras. Having captured

25% of the global market in digital cameras, Sony has expanded into developing periphery products, such as USB memory sticks. Given Sony's successful experiences, successful product design depends on the ability to cater to consumer demand by identifying the desired characteristics of products that appeal to certain population segments. For instance, with digital cameras, lightweight, handy and portable products equipped with large sized display screens are of priority concern. As digital cameras become common in daily use, manufacturers must incorporate appealing features and functions into product design.

範例 ii.

Fully understanding the market potential of the long term care sector in Taiwan has numerous benefits. As Taiwan's rapidly aging population reflects global trends, factors such as increasing daily pressures of modern living and family relations account for the significant rise in professional care providers for the disabled elderly. This trend in long term care reflects the need to enhance the quality of institutional-based healthcare island wide. This long-term care strategy focuses on small-scale healthcare institutes for several reasons. First, small-scale healthcare institutes are established more quickly and easier to manage than large-scale ones. Smaller-scale institutes enable us to more easily construct the infrastructure for a healthcare network, subsequently lowering operational costs in the long term and enhancing one's market share in the long term care sector. Second, market differentiation will enable smaller-scale institutes to avoid price wars with larger-scale ones that are supported by financial groups. Third, small-scale institutes focus on growth aspects that create lower financial risks. According to a recent market survey, although only around 20% of all disabled elderly in Taiwan receive institutional-based care, the market demand for institutional-based care is 30% (Department of Health, 1997). The 10% difference is equivalent to a market scale of at least 18,000 individuals. Moreover, the annual growth rate for disabled elderly in Taiwan is nearly 20 %.

範例 iii.

Taiwan's economy eventually reached full recovery following the SARS crisis of 2004, with economic indicators pointing towards sustained growth once again. Support from the World Bank helped to stimulate the domestic market, further contributing to intense competition within the financial sector. Since 1992, the Taiwanese government has devoted considerable resources to creating new banks. Given the excessive number of financial institutions that led to stringent competition, the quality of financial services deteriorated and the profit margin of the financial sector declined as well. To effectively cope with the global trend of liberalization and internationalization of financial services, the Taiwanese government enacted the Financial Holding Company Act to make the local financial sector more competitive. A notable example, credit cooperatives or banks have merged with other financial institutions to effectively respond to the influx of overseas institutions into the Taiwan market following the island's recent accession to the World Trade Organization. Although only thirty credit cooperatives are currently operating in Taiwan, some of them have been quite successful. Therefore, either mergers or the establishment of new banks are not the only means of corporate survival in the market. The financial sector in Taiwan should enhance its global competitiveness by fostering customer relations, enhancing business management practices and improving operational performance.

範例 iv.

Biotechnology impacts many aspects of daily life, including many foods and drugs consumed daily. Its applications are widespread - from agriculture to medicine - all capable of benefiting humans. This technology is commonly referred to as "The Industrial Revolution of the 21st Century". Given the increasing number of elderly and diabetic individuals, the market demand for curative products that can treat difficult-to-heal wounds is growing, as indicated by forecasted global market revenues to range from $US4,200,000,000 to 6,400,000,000 dollars by 2009. In particular, biopharmaceuticals play a major role in enhancing global production of biotechnology-related products owing to their heavy emphasis on target diseases, ability to provide more effective and potent action, few side effects and potential to cure diseases rather than merely treat the symptoms. With the Taiwanese Government having prioritized biotechnology as a major area of industrial development, this sector is widely anticipated to become a mainstream scientific discipline island wide. Although the global economy has not yet fully recovered, the medical treatment sector is still a growth industry, given continuous efforts to enhance the quality of life.

範例 v.

Natural materials are selected for bed mattresses to provide consumers with a comfortable posture for complete relaxation. Manufactured in Germany, these mattresses are imported to Taiwan. Natural materials of horse hair, wool and cotton comprise the mattress materials. Horse hair passes through a comb or falls by itself, followed by steaming, which creates more elasticity and ventilation. Wool passes through a clear current, thus avoiding chemical additives in its composition. Cotton is gathered in organic cotton fields, also without chemical additives in its composition, as well as insecticides or pesticide. Moreover, the cotton (once ripe) is picked by hand. Established in 1925, the German manufacturer has cooperated with its Taiwanese representative agent for an extended period, making it its general agent in Asia. A new strategy for providing the company with a larger presence in Asia is underway. First, five chain stores and five franchisee stores will be established in Taiwan, followed by establishment of a branch company in China. Prospects for success are optimistic given Asia' s increasing emphasis on environmentally friendly products. Therefore, the company' s emphasis on fulfilling consumer' s demand is widely anticipated to yield significant benefits. For the next two years, $20,000,000 will be invested in establishing five chain stores, along with an administrative department to determine the best location for those stores Additionally, an operating budget of $1,000,000 will be allocated. Before 2009, generated revenues will hopefully reach $5,000,000, with a branch company to be established in China by 2010.

B.產品或服務研發

1.產品開發的現階段狀況

- Despite the increasing number of solar energy construction projects in Taiwan, island wide development of this energy alternative is still in its preliminary phase, making it still impractical for satisfying daily consumption needs. Still, overseas solar energy firms have successfully transferred relevant technologies that will ultimately make Taiwan less dependent on non-renewable energy sources.

- Implementation of the National Health Insurance (NHI) scheme in 1995 has significantly impacted the quality of medical health care in several ways. Patients now enjoy an increasing number of options when selecting from a diverse array of medical service providers.

2.市場的評價

- Market growth has shifted from industrialized countries to developing ones in South America, Eastern Europe and Asia (especially China), with the compound annual growth rate exceeding 10.4% over the past five years.

- Given that Taiwan's economy has become more international-oriented following widespread market liberalization and privatization since 1990, the average annual income has skyrocketed to $US13,000 with 400,000 to 500,000 automobiles manufactured island wide.

3. 產品開發的獨創性和特色

- Recent surveys on consumer preferences in the digital camera market confer that lightweight, handy and portable products with a large size display screen are of priority concern. Such preferences provide a valuable reference for design specifications when Taiwanese manufacturers create innovative products.

- Taiwan's logistics sector integrates the activities of hundreds of companies in delivery, purchasing, bidding, distributing and transporting goods to retail outlets.

4. 台灣同類產品的主要製造廠商

- With Suzuki, Nissan, Formosa and Hyundai Motor Corporations taking the initiative in Taiwan, automobile manufacturing has approached a certain degree of maturity given the advanced production technologies adopted, variety in exterior and the latest product functions.

- Domestically, Run-tai Construction Company is collaborating with Chang Gung Memorial Hospital in establishing an elderly community based on the above principles. Given Taiwan's growing elderly population, elevated living standards, and preference of more elderly individuals to live independently of their children, the potential for this growth sector is enormous.

5.在既有領域下未來產品開發的原因

- Effective product design depends on consumer demand, which varies based on the unique characteristics of a certain target population segment. In sum, as digital cameras become a common household item, manufacturers must expand upon available functions to comply with an increasingly complex consumer demand.

- Continued growth of this sector largely depends on the ability of local manufacturers to continuously adopt new manufacturing technologies and offer variety in exterior features and digital products in the interior.

範例 i.

First adopted in Japan in 1995 and then in Taiwan in 2001 by First International Telecom, PHS mobile phones address the safety concerns of handset users with its use of a low electromagnetic wave, subsequently gaining a significant competitive edge in the local retail market. As the main competitor of PHS in the island's mobile phone market, Chunghwa Telecom Company, Taiwan Mobile Company and Far Eastone Telecommunications Company have gained a market niche through their sales promotional strategies that cater to customer needs to ensure flexible and reliable communication. Despite its lower market share in the mobile phone sector, PHS appears to have a lower mobile phone market. Still, PHS retailers strive to provide consumers with a low electromagnetic wave that addresses environmental and health concerns. Currently, Taiwan has only one manufacturer of mobile phones with a low electromagnetic wave, explaining its lower market share than other mobile phone providers. As for advertising strategy, PHS retailers heavily emphasize public education on the merits of mobile communications that are both convenient and environmentally friendly. For instance, as many hospitals discourage the use of medical instrumentation that contains a high electromagnetic wave interference, medical staff can only use pager devices, which is in contrast to the larger majority of handset users that own GSM mobile phones. With PHS's lower electromagnetic wave, medical staff can freely communicate with each other when an emergency arises. As the general public becomes environmentally conscious, PHS products and services will gain a greater market share in the local mobile communications sector.

範例 ii.

Living standards in Taiwan have dramatically increased in recent years, as evidenced by average per capita incomes surpassing $US 10,000 and the island's position among five of the largest holders of foreign exchange reserves worldwide. As the Taiwanese economy prospers, a diverse array of industrial fields is burgeoning as well. Of the advanced medical care treatments available, the cosmetic surgery sector is a notable example. The number of cosmetic surgeries and treatments increased five folds over a one-year period (2003-2004): from approximately 30,000 to more than 150,000. The strong demand for cosmetic surgery closely corresponds not only to the public perception that an attractive appearance is vital for professional and social settings, but also to the beliefs of many Taiwanese that one's face or body can influence one's fate, as postulated by Chinese geomantic theory. While recent studies show that both genders are equally attracted to cosmetic surgery, women over 40 years old account for 70% of all treatments. Improvements in one's face and overall appearance instill in clients a greater sense of self-confidence. While all surgical procedures are administered by licensed physicians, a wide array of cosmetic products to maintain one's appearance is available.

範例 iii.

Facial mask products for cosmetic purposes have been extremely popular in Taiwan for many years, with a sales volume of around 120 million units so far. As Taiwanese women become more involved in athletics, the potential for further market expansion appears to be limitless. In purchasing these products, consumers are concerned with brand recognition, healthy ingredients and a competitive retail price. Given the strong sunlight in Taiwan, especially during increasingly longer summer periods, Taiwanese women generally pay much attention to the whitening of their skin, as evidenced by the 15 million facial masks sold annually. More than the whitening effect, Taiwanese women also enjoy facial masks for their ability to relax the mind and nourish one's skin through proper treatment. The large number of skin care products available reflects the priority that Taiwanese women place on daily use of facial masks for whitening and restorative purposes — regardless of whether they are adolescents, housewives, office employees or administrative personnel. In this market, 25-45 year old women account for 70% of all consumers. Additionally, various technologies are adopted in developing advanced face masks that whiten skin, maintain facial moisture, profess anti-ageing capabilities and contain collagen. Given these features, facial masks can be purchased at a relatively low retail price. Further market potential can be found in the use of facial mask products among men. Other than facial masks, similar products are available for breasts, the neck and hands.

範例 iv.

The intensely competitive information technology sector has significantly decreased the time to market delivery of advanced digital cameras. A strong product brand image that distinguishes itself from other digital cameras in terms of superior features or functions is essential for commercial success. Recent surveys on consumer preferences in the digital camera market confer that lightweight, handy and portable products with a large size display screen are of priority concern. Such preferences provide a valuable reference for design specifications when Taiwanese manufacturers create innovative products. When purchasing cameras, most consumers consider pixel hues, overall appearance and operating interface smoothness, areas in which Sony has excelled in product innovation. Among the many digital camera manufacturers in Taiwan, BenQ, Premier and Acer also strive for creativity in product innovation. As consumer demand for digital cameras rapidly increases owing to advanced image technologies and competitive retail prices, Sony aspires to maintain its leading market share of 25% of the global market by sustaining a certain price elasticity and expanding the functions of its digital cameras. Sony is also developing peripheral products such as USB memory sticks to build upon its competitive edge in the digital camera sector. Importantly, effective product design depends on consumer demand, which varies based on the unique characteristics of a certain target population segment. In sum, as digital cameras become a common household item, manufacturers must expand upon available functions to comply with an increasingly complex consumer demand.

範例 v.

Implementation of the National Health Insurance (NHI) scheme in 1995 has significantly impacted the quality of medical health care in several ways. Patients now enjoy an increasing number of options when selecting from a diverse array of medical service providers. With growing public concern over patients' rights, individuals are encouraged to become more responsible for decisions affecting their health. In the recent decade, North American and European countries have aggressively accumulated and released patient-related data to assess the performances of healthcare providers. Such a trend reflects the importance of consumer-oriented marketing in forecasting trends in usage of healthcare services. As early as 1983, NHI adopted the consumer health care decision model to analyze questionnaire results from the American public so that healthcare marketing strategies could understand consumer patterns in selecting medical care and ultimately enhance the reputation of practicing physicians. While a study in 1999 demonstrated that attracting a new customer would cost a company ten times as much as keeping an old one, in that same year, Taylor and Cosenza adopted the hospital shopping model choice model (HSMCM), indicating that volatile environmental changes have led to the closure of many hospitals and clinics. Patients must therefore carefully scrutinize all available medical resources before reaching a decision. To do so, the rapid circulation of online information and promotional materials makes it easier for consumers to make the best choice, ultimately increasing the efficiency of the healthcare sector. While the NHI scheme has allowed Taiwanese residents to freely choose their healthcare providers, whether they have sufficient information to distinguish between those healthcare providers is unclear. More than the proximity of the hospital to the individual's residence, other factors must be included, such as patient expenses not covered by NHI. Clearly identifying those factors that influence the medical consumer's decision as to which healthcare provider to choose can help healthcare practitioners to evaluate the effectiveness of their offered medical services.

C. 專案描述

1.專案的背景

The fiercely competitive management environment has necessitated that enterprises initiate many innovative plans, including organizational restructuring or personnel reductions. Likewise, outsourcing is a major industrial trend aimed at ensuring corporate survival.

Given changes in the domestic market, a global recession in recent years and Taiwan's recent entry into the World Trade Organization, the local automotive sector has changed its business management practices accordingly. When launching a new model in the intensely competitive market, automotive companies strive to efficiently use available resources and determine which marketing strategies are most effective in stimulating consumer purchasing.

2.市場的訊息

With consumer demand expected to grow continuously, according to expert forecasts, solar energy use will increase at a rate of 33%, with its output value targeted at $US 1,400,000,000.
According to statistics, acquiring a new customer costs at least five to nine times more than maintaining an old one. Additionally, increasing customer loyalty by 5% can elevate profits by 25-85%.

3.專案的目標

More than an alternative to petroleum, as an unlimited power source that does not extract an environmental cost, solar energy represents a growing environmental consciousness aimed at protecting the world's natural resources.

The government should encourage this initiative in two ways. Legislation governing health food standards should be implemented, as is already done globally. Second, efforts should be directed towards strengthening the production capacity of the health food sector locally because the mostly imported health food products currently available do not necessarily suit local tastes and consumer demand.

4.專案的重要特色：專案的策略

Under the BOT scheme, the Taiwanese government initially holds the property rights of an infrastructure project and then transfers them to the private sector. The private sector is then responsible for investing in the project and implementing it for an agreed upon period. Once the period expires, the private sector transfers the project assets to the government with or without compensation.

Given the consolidation of construction personnel and subsequent increased workload in the civil engineering sector, project planners strive to effectively manage human resources as flexibly as possible. As management adopts various outsourcing strategies to alleviate the above predicament, whether outsourcing will extend to construction personnel to control expenditures and better allocate human resources remains to be seen.

5.專案的動向

While successful development of Taiwan's hi-tech industry is reflected in the Hsinchu Science-based Industrial Park and the Tainan Science Park, the new transportation network will further accelerate access and technological progress between the northern and southern parts of the island.

Despite the considerable time and capital investment required to successfully implement the above two strategies, doing so would not only encourage local manufacturers to opt for environmentally friendly alternatives, but also fill in the gap that is largely controlled by overseas producers.

範例 i.

Homemade style cookies in Taiwan are only available in coffee shops and convenience stores. To fully realize the market potential, local producers are aiming to expand to other Asian countries. Our recent undertaking has been to first establish several shops in Taiwan, with coffee kiosks located inside to encourage purchases of homemade style cookies. If successful, thirty retail outlets will be established in China and Japan. To satisfy a broad spectrum of consumer tastes and demand for certain flavors, producers will develop an extensive product line of homemade style cookies and other snack varieties, including bread, cakes and candy. Substantial investment must also be made in machinery and preservative containers to ensure that the cookies maintain their freshness and avoid the use of unhealthy additives that contribute to obesity. Although homemade style cookies are perishable items unable to maintain their freshness for an extended period, using fresh ingredients and avoiding the use of artificial additives can sustain product life. While our company accounts for 10% of all homemade style cookies sold in Taiwan, the above strategy could considerable raise product revenues. Hopefully, expanding the number of retail outlets island wide and exporting our products overseas will increase the company's market share to 30%. Selecting appropriate locations for such outlets and forming alliances with other cookie and beverage retailers are essential to project success. Finally, a marketing management department will be set up in the company to coordinate the activities of local and overseas retail operations, thus ensuring that all new product promotions can be implemented smoothly.

範例 ii.

Marketing of home massage appliances in Taiwan has become increasingly aggressive, as evidenced by the growing number of product brands and diverse product lines, elevated living standards island wide, high consumer expectations for quality merchandise, competitive retail prices and comprehensive customer service. Taiwan's accession to the World Trade Organization in 2002 created further marketing opportunities for luxury items such as these. Daily pressures of living, sedentary lifestyles owing to extensive use of computers, hectic work schedules, poor postures and lack of exercise have all contributed to shoulder aches and pain in the loins, back and other parts of the body. As a leading manufacturer of home massage appliances, Taiwan OSIM has patented several of its product technologies and distinguishes itself from other manufacturers in its extensive use of color and various materials such as magnetite in its products. Still, increasing sales revenues in this market do not necessarily imply a stronger purchasing power of consumers. To increase its consumer appeal, Taiwan OSIM has developed an extensive product line that includes massage chairs, foot massagers, air massagers, slim belts, mattresses, pillows and related accessories. OSIM's marketing strategy is spearheaded in three directions. First, healthy living implies that modern lifestyles can be exciting and place a strain on our bodies, which require solace. Second, relaxation is similar to a soothing balm for one's body and spirit. OSIM has innovatively integrated several product technologies to ensure relaxing comfort. Massage chairs, hand held massagers and foot soothers are among its top selling products. Third, beauty is a result of invigorated energy that comes through relaxation that OSIM's products offer. For instance, OSIM's slim belt comes in a variety of product specifications so that users can trim their waistlines, provide back support, relieve back strain with uniquely installed biomagnets and improve one's posture effortlessly. Given the above marketing emphasis, Taiwan OSIM has exported its products overseas, having been warmly received in countries such as Australia, South Africa and the United States. Product revenues this year are expected to increase by 35% over profits in 2004.

範例 iii.

Enterprise resource planning (ERP) refers to the efficient planning of an enterprise's resources through the integrated use of information science and technology. Comprising the financial affairs section of an enterprise, accountants, manufacturers, suppliers and administrative managers, ERP strives to integrate human resources in a synergistic manner. Merging the workflow of various departments can enable managers to make decisions efficiently based on available operations-related information, thus improving both management efficiency and resource utilization. To enhance emergency unit procedures in hospitals, an ERP-based project has been adopted worldwide (including Germany and Japan) to upgrade the quality of medical services, with fruitful results generated so far. Although extensively adopted in many business circles, the ERP system is seldom adopted in the medical sector. Our current project attempts to adopt the ERP system in a clinical setting to enhance medical treatment. While focusing on patients receiving artificial appendages and individuals suffering from hemorrhoids, this system can hopefully increase revenues by 30% and customer satisfaction with medical quality from70% to 85%. An effective means of achieving these goals is to evaluate the quality of emergency medical procedures from the perspective of consumer preferences, market data, standard operating procedures in a hospital and current financial situation. Enhanced customer satisfaction will ultimately improve the quality of services provided by medical personnel, streamline workflow procedures and simplify financial procedures. This unique ERP system-based project will greatly benefit emergency and surgery procedures by adhering to well established goals and procedures. Doing so will increase satisfaction not only among patients, but hospital staff members as well.

範例 iv.

Business management practices of the domestic logistics sector have dramatically transformed local companies given Taiwan's recent entry into the World Trade Organization. While encompassing wholesale outlets, warehouse and transportation sectors, the logistics market attempts to fill the demand for common goods (e.g., books, clothing, electrical appliances and medicine), specialized goods, cargo delivery from storage, product distribution, integration of information systems, payment of goods, as well as delivery and accounting practices. The information system that the logistics sector adopts generally focuses on accounting concerns of controlling stock inventory and the outflow of goods. Software engineers spend considerable effort in reducing to a minimum the number of personnel required in the entire logistics process: from selecting the optimal delivery route to choosing the most effective arrangement of goods for delivery. Given their rapid development domestically, logistics centers must effectively utilize technological know how in their management practices, often having to develop their own software to outsource their software needs to contract workers — all while maintaining the confidentiality of their business practices. Therefore, each logistics company has its own unique management style, such as in analyzing current market situations, handling customer-related issues and managing the turnover rate of goods as flexibly as possible. To satisfy the information technology needs of a logistics company, its software must comply with corporate standards and adjust correspondingly to personnel requirements. Specifically, its accounting line system must reflect the software's capability to integrate all available data that enables the company's administrators to make the most effective decisions in daily operations.

範例 v.

Both profit and non-profit organizations devote considerable resources to understanding and satisfying consumer demand. According to statistics, acquiring a new customer costs at least five to nine times more than maintaining an old one.

Additionally, increasing customer loyalty by 5% can elevate profits by 25-85%. Therefore, while striving to satisfy the needs of consumers and increase their loyalty, enterprises hope to ultimately elevate corporate profits. With the advent of Taiwan's National Health Insurance System, the medical care sector has become fiercely competitive given increasingly limited governmental subsidies for medical treatment and services. Healthcare organizations have thus begun to focus on how to upgrade their management practices, as evidenced by the increasing use of marketing and customer management practices in daily operations. Medical organizations have also been reexamining their public image, quality of services, as well as customer satisfaction and loyalty. Whereas the manufacturing sector has traditionally emphasized customer satisfaction and loyalty in its management approach, these areas have seldom been explored in the medical sector. Therefore, the project that I am currently involved in attempts to determine levels of customer satisfaction and loyalty towards medical services by adopting a cross-sectional method that combines the American Customer Satisfaction Index (ACSI) and the European Customer Satisfaction Index (ECSI) and considering the following areas: company image, consumer expectations, consumer attitudes, quality of service, quantitative value of consumer attitudes, customer satisfaction, customer complaints and customer loyalty. By covering patients in five Hsinchu area hospitals, the study adopts medical marketing research methods and a structured survey questionnaire to evaluate factors influencing customer satisfaction and loyalty. Quantitative results of customer satisfaction and loyalty will hopefully contribute to efforts to enhance the quality of medical services, providing a valuable reference for hospital administrators attempting to increase their competitiveness in daily operations.

D. 公司或組織介紹

1.簡要陳述組織所屬之產業的概況

Elderly individuals in Taiwan over 65 years old account for 9.4% of the total population, surpassing the United Nations' definition of an aging society. Given this trend, market opportunities for senior citizen residential communities island wide are estimated at more than $US 100,000,000 annually.

Recent efforts within the printing sector to integrate graphic communications and computer technology have led to the extensive use of digitalized color images in graphic communications for areas such as Internet-based applications, new printing materials and media applications.

2.組織的使命；組織的發展沿革

Whereas older correctional facilities focused on isolating inmates from the general population without much concern for their rehabilitation, modern correctional facilities adopt a more humanistic approach towards orienting and reforming the incarcerated so that they will have practical skills upon re-entering society.

Established in 1997 and having evolved into a leading global supplier of computers for industrial use, IEI Technology Corporation wholeheartedly strives to achieve customer satisfaction through its product technology research, design and commercialization, as well as flexible manufacturing systems, marketing, sales and customer service.

3.組織的架構

While the Taiwan headquarters consists of computer, printing, business and accounting departments, the Chinese branch comprises computer, printing, quality control, processing, business and accounting departments

The six bureaus that operate with National Health Insurance process insurance applications receive insurance premiums, audit medical expenses and consult with hospitals on how to enhance their administrative management practices. Of the twenty four liaison offices located conveniently island wide to continuously upgrade the quality of medical services, the National Health Insurance Bureau staffed 2,536 employees and 534 contract workers in 2004.

4.組織最新的科技成就

Regardless of the urgency of a customer's query regarding application needs or trouble-shooting assistance, IEI currently replies to 93% of incoming queries related to sales within 12 hours and queries related to technical support within 24 hours. Finally, to ensure on-time delivery, IEI has successfully established a highly integrated supply chain and logistics system to allocate key components efficiently in order to increase manufacturing efficiency.

As semiconductor manufacturing technologies have entered the nano era, National Nano Device Laboratories (NDL) closely collaborates with academic institutions and private enterprises to remain abreast of the latest advances in this challenging field. Specifically, the emergence of new generation nano materials and applications reflects a global trend in semiconductor manufacturing technologies that requires extensive knowledge of multidisciplinary fields, a long term commitment and international exchange opportunities.

5.結論 (未來發展方向)

As for future directions, with governmental initiatives to liberalize and internationalize the local economy, TSC will eventually become a privatize enterprise and, in doing so, expose itself to greater competition overseas.

Strengthened by technological transfers of expertise from abroad and rich management experience, RUENTEX remains optimistic about its future prospects in the elderly housing market island wide.

範例 i.

Established in 1979, How-Well Enterprise strives to combat the increasing environmental loading that Taiwan faces given strong public awareness over environmental protection issues and stringent governmental legislation. While adopting state-of-the-art Japanese waste treatment technologies, How-Well treats medical waste efficiently and safely through an incinerator approach that heavily stresses pollution prevention measures, as evidenced by its ISO-1400 accreditation in 1998. Given the shortage of medical waste treatment and incineration facilities, How-Well has accumulated many years of practical experience in incinerating medical waste through state-of-the-art pollution prevention facilities. As the private sector in this area rapidly evolves, How-Well develops innovative approaches to planning and design in resolving the seemingly endless amount of generated waste. Professional tasks in this area include the following: incinerator design and manufacturing procedures, standard operational procedures of incinerators and commercial processes involved in removing medical waste. Among the areas that How-Well services include factories, hospitals, schools, hotels, amusement parks and airports.

範例 ii.

Established in 1997 and having evolved into a leading global supplier of computers for industrial use, IEI Technology Corporation wholeheartedly strives to achieve customer satisfaction through its product technology research, design and commercialization, as well as flexible manufacturing systems, marketing, sales and customer service. As evidence of its success in the computer market, IEI has successfully commercialized over 600 products, including single board computers, servers, industrial PC chassis, workstations, panel PCs, flash disks, PC/104 products, power supply units and backplanes. IEI's products can be found in a diverse array of computer-based applications, including factory automation, voice over the network telephone services, networking appliances, security systems, POS systems, national defense, police administration, transportation, communication base stations and medical instrumentation. To fully satisfy customers' diversified requirements, IEI has heavily invested in flexible manufacturing systems that emphasize quality production and timely delivery with minimal cost. ICP’ s two manufacturing facilities cover more than 139,117 square feet. Additionally, ICP has also heavily invested in automatic insertion equipment, including five advanced SMD assembly lines, five AOI automatic inspection devices, two DIP assembly lines and an EMI laboratory. Research and Development forms the basis of IEI innovation, with more than eighty technical experts in the R&D Division, which receives a large proportion of the company’ s annual budget. IEI's R&D Division comprises the following groups: single board computers, system peripheral devices, Internet products, product management, design quality verification and the newly established SOC team. All research groups have strong technical design capabilities and innovativeness in product technology development. As for customer service, IEI’ s Sales Division provides 24 hour online customer support. Regardless of the urgency of a customer’ s query regarding application needs or trouble-shooting assistance, IEI currently replies to 93% of incoming queries related to sales within 12 hours and queries related to technical support within 24 hours. Finally, to ensure on-time delivery, IEI has successfully established a highly integrated supply chain and logistics system to allocate key components efficiently in order to increase manufacturing efficiency. Armed with build-to-order (BTO) capabilities, IEI's logistics center can ensure timely product delivery based on a client' s customized needs.

範例 iii.

Besides the Blockbuster franchise, Cine-Asia Entertainment (CAE) Company is a major provider of DVD and VCD rentals for home use in Taiwan. Given widespread video piracy and copyright infringements, these two companies had difficulty in turning a profit in the early stages. Previously accustomed to ignoring copyright laws, many traditional movie rental stores could not adjust, subsequently going out of business given high operating costs associated with leasing copyrighted movies for rental purposes. Established in 1999, CAE Company maintains more than 100 franchises island wide. In addition to establishing a nationwide network of audio-visual service providers and enhancing the quality of home entertainment, CAE Company strives to provide a diverse array of audio-visual services and multimedia products, foster a warm climate of trust with customers and adopt the latest innovations in customer service As a store manager in the retail sales department, I was responsible for managing all aspects of daily operations, such as estimating the quantity of DVDs and VCDs to be stocked in inventory. In addition to orienting staff on how to promote products among customers, store managers must offer their recommendations to the CAE Company headquarters on the most effective marketing strategies to adopt. Exposure to this environment familiarized me with many promotional strategies on how to increase a company' s competitiveness.

範例 iv.

Established in 1990, Real-Sun Information Technology Company has nine branch offices island wide in Taipei, Taoyuan, Taichung, Jiayi, Tainan, Kaohsiung, Pingtung, Yilan and Hualien. The administrative staff and service engineers at Real-Sun come from a diverse array of multi-disciplinary fields, each having anywhere from three to twelve years of professional experience with medical-oriented computer systems. Renowned for its ability to offer 24 hour online technical service support and on-site maintenance within four hours after a problem is reported, Real-Sun services more than 5,000 physicians and 8,000 nurses through its medical information systems and offers eleven areas of computerized services. The company's medical information-oriented software can easily adapt to a physician's routine and individual habits, as well as integrate into a permanent system, as evidenced by the thirty six local hospitals and 2,800 clinics currently using the company's designed system. Of particular interest is Real-Sun's computerized system for professional clinics and hospitals, which is a software package for projects under development. The software package can be used to register patients, process outpatient services, levy medical expenses, apply for governmental subsidy of medical fees and make drug prescriptions. Additionally, the system interface is easily operated, with initial learning requiring an average of 30 minutes. Adopting this computerized system has three main features: wide frequency networks are increasingly popular; software providers and users can establish a reliable VPN enterprise network; and software providers must have a basic technological know how of how the service network operates. Besides conforming to the above features, Real-Sun's system has already been successfully integrated into PACS, LIS, and HL7 operations, ensuring that medical software specifications comply with user demands.

範例 v.

Established nearly seven decades ago, the Third Credit Cooperative Bank of Hsinchu currently has a stock value of roughly $NT 890,000,000. Of the company's 200 employees, more than half have acquired at least a bachelor's degree; in addition, the average employee is 35 years old. Among the fourteen other lending agencies in Hsinchu, the Third Credit Cooperative of Hsinchu is of medium size. The Third Credit Cooperative offers services similar to those of other banks, including various schemes to save money as well as provide short-term and long-term loans. With the company organized into the banking department, accounting department, information department, marketing department and general affairs department, each department strives to remain productive in order to serve customers fully. Market liberalization policies of the Taiwanese government since 1992 have led to the emergence of several banks, with financial organizations springing up quickly and making the finance market increasingly competitive. The Third Credit Cooperative of Hsinchu thus exerts considerable effort in maintaining employee morale, providing efficient banking services and reducing investment risk to secure long term corporate profits.

E.組或部門介紹

1.介紹部門所屬的組織或公司

Although not directly generating revenues for the company, the Programming Division still plays an indispensable role in daily operations by integrating the efforts of other divisions to execute a cohesive business strategy based on customer relations management-related practices (CRM) and other marketing strategies.

Within the company, the Enterprise Planning Department, Recycling Department,

Environmental Engineering Department and Engineering Maintenance Department handle daily operations, as staffed by ninety employees. Responsible for handling waste efficiently and safely, the Recycling Department treats infectious waste, waste from the steel industry and heavy metals.

2.部門的組織架構

The regional National Health Insurance Bureau consists of seven sections and four offices: first underwriting section, second underwriting section, third underwriting section, medical affairs section, outpatient expenditure section, inpatient expenditure section, medical expenditure review, personnel office, information management office, civil service ethics office and accounting office - all of which are under the direction of the General Secretariat.

As a leading cord blood bank in Taiwan with the safest cord blood cryopreservation facility in Asia, Sino Cell Technologies provides high quality cord blood stem cell processing and storage capabilities. Within the organization, the Cord Blood Registry processes, tests and stores cord blood stem cells, assuming responsibility for processing, quality control and quality-assurance metrics that comply with FDA guidelines.

3.部門的人才來源和教育背景

Headed by a director, the department comprises two senior engineers and three operators, with each acquiring at least a bachelor' s degree and an average of five years professional experience.

Comprising a departmental head, four civil engineers, five technicians and an administrative assistant, 75% of the departmental staff have acquired at least a bachelor's degree in a construction engineering-related field. Recruited through the human resources division via newspaper advertisements and a rigorous interview process, departmental staff adopts a professional and congenial attitude towards serving its clients and the general public.

4.部門的使命

While devoting its energies to devising and implementing successful civil engineering projects, as well as coordinating governmental-sponsored projects, the Programming Division strives to give the company a competitive edge in the housing construction sector by adopting sound management practices that will ensure customer satisfaction.

The Department is responsible for devising information services, suggesting how to adopt new information technologies, analyzing the feasibility of adopting new product technologies, coordinating departmental training, establishing departmental incentives and regulations, assigning personnel to service various divisions within the company, coordinating activities of the information technology department with those of other departments and outsourcing IT needs to contracted information technology vendors.

5.部門之製造或研究的能力

The Department adopts advanced technological equipment imported from Japan that melts infectious waste at high temperatures. In this process, infectious waste is collected and then treated at combustion temperatures ranging from 1650℃ ~3000℃. By adopting this procedure, the Recycling Department services hospitals, schools, airports, hotels and community disposal facilities.

Committed to high quality printing, the Computer Department has state-of-the-art equipment to complete the task at hand, including a comprehensive pre-press desktop publishing system a color plate manufacturing system based on Apple Computer, color scanners, laser image output, laser printers, film copying machines, film developers, a plate copying apparatus and draft printers.

6.部門提供的產業服務

The Bureau of Monetary Affairs is divided into six units to handle different aspects of daily operations: monitoring the development of financial organizations, making relevant regulations, supervising financial market reforms, addressing stock-related problems, analyzing financial events and facilitating mergers of financial organizations.

The Healthcare Center takes full advantage of its numerous years of solid experience by focusing on the health concerns of the elderly, implementing health sustaining programs, providing individual counseling on how to effectively address the physiological and psychological needs of the elderly and, ultimately, establishing a sustainable healthcare system that caters to the special needs of Taiwan' s aging population.

範例 i.

Established in 1988, Jia de Technological Development Company treats waste according to stringent quality control measures in order to avert environmental pollution. Within the company, the Enterprise Planning Department, Recycling Department, Environmental Engineering Department and Engineering Maintenance Department handle daily operations, as staffed by 90 employees. Responsible for handling waste efficiently and safely, the Recycling Department treats infectious waste, waste from the steel industry and heavy metals. Additionally, the Department adopts advanced technological equipment imported from Japan that melts infectious waste at high temperatures. In this process, infectious waste is collected and then treated at combustion temperatures ranging from 1650℃~3000℃. Successfully implemented in Japan for more than a decade, this treatment procedure ensures the safety of medical personnel and elimination of infectious waste. By adopting this procedure, the Recycling Department services hospitals, schools, airports, hotels and community disposal facilities.

範例 ii.

The Respiratory Care Center at Taichung's Shun-Tian General Hospital is renowned for its treatment of chronic disease patients in central Taiwan. Advances in medical treatment and an increasing number of patients have led to an expansion of respiratory care facilities and an increase in bed capacity. Given that patients in the Respiratory Care Center require professional care and round-the-clock attention, Shun-Tian General Hospital recently renovated its medical facilities, with special attention paid to the Radiology Department. Given its pivotal role in the hospital's daily operations, the Radiology Department runs a general radiology area with portable instrumentation and four x-ray examination rooms that take the following: a) routine x-ray images of specific areas such as the chest, abdomen, skull, limbs and spine; b) unique x-ray images such as the lower and upper GI; c) computed tomography images for head, abdominal and spinal scans which are common for the emergency care department and the orthopedics department; and d) ultrasound images for patients requiring sonar scans of the hand and abdomen. Moreover, given rapid changes in the medical market, the radiology department has effectively responded to the need to take patients' medical images from portable equipment.

範例 iii.

Wang King Color Printing Company consists of the Computer, Printing, Business and Accounting Departments. Staffed by 23 employees, the company strives to achieve a daily operational flow that runs an efficient, automated working environment that can increase production efficiency with accuracy and precision. A notable example is the Computer Department, which contains four employees with an average work experience of over three years in the printing sector. Committed to high quality printing, the Computer Department has state-of-the-art equipment to complete the task at hand, including a comprehensive pre-press desktop publishing system a color plate manufacturing system based on Apple Computer, color scanners, laser image output, laser printers, film copying machines, film developers, a plate copying apparatus and draft printers. While adhering to the motto that design is the visual communication of information, the Department adopts a variety of skills in creating hi quality prints. Additionally, the Department specializes in customizing color prints according to consumer preferences. Besides pursuing commercial and creativity-related interests, the Department heavily stresses a professional attitude towards serving customers as comprehensively as possible. Equipped with a competitive spirit, solid management experience and advanced technological capabilities, the Computer Department prides itself in whatever printed products it creates, regardless of whether such products are name cards, books, packaging boxes, tapes, labels, calendars, posters, booklets or catalogues.

範例 iv.

The Cine-Asia Entertainment Company in Taiwan is a home entertainment franchise that rents DVDs and VCDs, with more than 100 branches located island wide. Having significantly contributed to the island’s service sector, the company comprises an accounting department, finance department, franchisee management department, purchasing department, logistics department, information management department, marketing department, as well as public relations and general affairs department. A manager runs the daily operations of each department, which must operate efficiently and productively. A notable example is the franchisee management department, which supervises the operations of more than 100 retail outlets. Responsibilities at the retail outlet level include ordering and returning entertainment products, forecasting sales, communicating effectively and coordinating the efforts of staff in the retail outlets. As the franchisee management department often involves coming into contact with customers at the retail outlet level, staff must enthusiastically engage with local staff with a strong sense of commitment. Specifically, the department strives to generate company revenues, promote the entertainment quality of company products, instill in customers a sense of concern and respect for their particular viewing or listening tastes and foster a harmonious home environment with the latest DVD or VCD movie releases. Customer satisfaction will ultimately yield company profits, giving CAE a competitive edge in the intensely competitive home entertainment sector.

範例 v.

The Blood Inspection Department that I belong to at Blessed Virgin Mary Hospital as a medical technologist has nineteen staff members, including a general director, seventeen medical technologists, a laboratory assistant and an attending physician who serves as a consultant for the blood bank. All staff members have attained at least a master's degree. The department consists of the following six working groups: biochemical, blood work, serum, blood bank, blood examination and bacterium. Departmental work space is divided into three areas: the hospital' s comprehensive examination room, outpatient services and an emergency unit. As for the department' s instrumentation, each group adopts state-of-the-art automation instrumentation to facilitate all blood-related inspections in the hospital. Additionally, the department heavily emphasizes continuous training as an effective means of constantly upgrading staff skills. For instance, on-the一job-training is held each Thursday in the form of seminars held by eminent medical technologists from other institutes to share their expertise. Moreover, representatives from each working group are often selected to receive training outside of the hospital. A well-stocked hospital library with the latest medical periodicals and magazines keeps the staff abreast of the latest trends in their respective fields. As evidence of the department' s commitment to excellence, the National Institute of Health (NIH) selected one of the department' s clinical laboratories as an advisor for similar laboratories conducting bacterium examinations. Having received several commendations from the NIH as well as local and overseas medical associations for its exemplary operations, the department undertakes many parasitic examinations, especially given the large number of migrant laborers that come to the Blessed Virgin Mary Hospital for inspection.

F.工業介紹

1.此種產業在台灣的一般特色

While Taiwan has focused on developing the semiconductor and OEM industries in recent decades, the question arises as from where the island' s latest technological advancements will emerge. Restated, exactly where does Taiwan rank in the globalization scheme? With the rapidly elderly population worldwide, the long term health sector will undoubtedly emerge as a leading industry.

Characterized as low polluting and capital intensive, the biotechnology industry employs highly qualified personnel from multidisciplinary fields and develops product technologies with a high economic return. Given those characteristics, R&D budget accounts for a large proportion of the company' s operating budget, which is understandable if a company wishes to enhance its competitiveness.

2.產業所面臨的困境

Among the several challenges that the automotive industry in Taiwan has faced include the extended time required for automobile design, subsequently increasing new product delivery time to the market; large control that the overseas headquarters maintains in which technologies to adopt and how original models should be modified, thus limiting the design capacity of the local manufacturer; and inability of the overseas headquarters to directly promote the competitiveness of its products in the local market sector besides the quality of the car brand itself.

A rapidly aging population, advances in medical technology that increase average life expectancy and the increasing availability of medical services have contributed to a situation in which increased medical fee expenditures exceed the amount of National Health Insurance premiums collected.

3.簡述一或二個目前活動的重點

The paper processing industry significantly contributes to domestic and overseas forestation efforts. For instance, Yuen Foong Yu Paper Corporation and the Chinese Pulp Corporation have collaborated in forestation of 2,000 hectares in Vietnam, with Yuen Chi Paper Corporation investing in 150,000 hectares of forestation in Indonesia.

The Taiwanese government has recently promoted a new policy entitled "Broaden the sources of National Health Insurance generated income to reduce medical expenditures" as an incentive for hospital administrators to increase productivity and efficiency, as well as assure the general public of the continued provision of high quality medical care.

4.產業採用的科技

In terms of the development of GPS navigational systems, domestic manufacturer capacity is nearly equivalent to that of overseas automakers although locally developed automotive security systems and wireless on-line network capabilities still lag behind overseas counterparts. Given rapid scientific and technological advances, the automotive industry will heavily emphasize wireless on-line network capabilities, GPS navigational features and foolproof security systems.

To address this concern, several semiconductor manufacturers have recently entered this burgeoning market by introducing advanced manufacturing capabilities at a more competitive retail cost given the long-term investment involved in operating and maintaining this system.

5.台灣相關的研究發展設備

While requiring highly skilled personnel to advance biotechnology product applications, biomedical companies are especially concerned with the shortage of qualified staff with adequate skills to remain abreast of the latest developments. Education, training and information alone are insufficient to keep Taiwanese biotechnology firms in pace with technology-driven changes in the workplace. Therefore, in addition to enhanced training of research and managerial personnel in specific fields, local biotechnology and medical organizations are exploring ways to integrate other innovative fields such as bioinformatics and genomics into their research scope and product line.

Among the numerous new knowledge economy-oriented ventures underway, nanotechnology plays a leading role in scientific and technology-related research. For instance, after Taiwan University Research Group (TURG) commercialized its nanoparticles product, many local enterprises followed suit in entering the nanotechnology market, including the Industrial Technology Research Institute and the Textile Technology Research Institute. As the local nano-paint solvent manufacturing sector continues to expand, many industries have expressed interest in expanding their product lines to this area. Ultimately, R&D activities will determine whether the domestic nano-paint solvent manufacturing sector can compete globally.

範例 i.

Medical imagery belongs to the larger medical technology sector. While Taiwan has focused on developing the semiconductor and OEM industries in recent decades, the question arises as from where the island’s latest technological advancements will emerge. Restated, exactly where does Taiwan rank in the globalization scheme? With the rapidly elderly population worldwide, the long term health sector will undoubtedly emerge as a leading industry. The medical imagery sector in Taiwan is still in its developmental stage. The emergence of the biotechnology trend worldwide reflects the increasing emphasis on health, as evidenced by an increasing elderly population globally. Regardless of whether materials or the human body is concerned, technology advances from the macro to micro level are increasing, e.g., nanometer technology. For human eyes, researching a phenomenon sized at the nanolevel is nearly impossible. However, auxiliary instruments can facilitate research into molecular-sized objects. The computer enables such an investigation. For instance, medical imagery is placed in a human organ microscopically and then investigated at a molecular size. I prefer researching topics related to nuclear medicine as an excellent example of what molecular medicine can (entail or include). The medical imagery sector belongs to the long-term health care sector, an area which will undoubtedly play a pivotal role in science in the new century.

範例 ii.

The biotechnology sector in Taiwan encompasses a diverse range of multidisciplinary fields and closely collaborates with other industrial partners in exploring topics that would further elevate human living standards such as medical science. Development of this sector is still in its preliminary stage given the considerable investments still to be made in biological, scientific and technological instrumentation and facilities, as well as materials and subjects for experimentation, e.g., animals, plants and biomedicine components. DNA and genetic technologies are among the various fields explored. Notable examples of domestic institutions involved in related research include Academia Sinica, Industrial Technology Research Institute, Animal Technology Research Institute, clinical departments in hospitals, universities and R&D departments of private enterprises. Topics of particular concern are genetic research-related topics, including a genetic link in hereditary diseases, protein research, fermentation technology applications, in vitro research of cancer cells, healthcare research and food science-related topics, e.g., genetically modified foods and genetically modified organisms. Ultimately, above research efforts reflect the potential of biotechnology in elevating human living standards through breakthroughs in medical science and making life more convenient.

範例 iii.

The paper processing industry has expanded rapidly in the new century. Locally, Yuen Foong Yu Paper Corporation has significantly contributed to the development and supply of paper-related products in Taiwan. Paper factories recycle domestic wastepaper, yielding tremendous environmental benefits through garbage removal. Paper factories recycle and reuse roughly 2,800,000 thousand tons of domestically produced wastepaper annually. Comparatively, nine incinerators can only treat 2,7000 tons monthly, further reflecting the environmentally friendliness of the paper processing industry. According to the Industrial Technology Research Institute, one hectare of forested land can absorb 13 tons of carbon dioxide, yet further demonstrating the positive environmental impact that domestic paper factories have. Moreover, the paper processing industry significantly contributes to domestic and overseas forestation efforts. For instance, Yuen Foong Yu Paper Corporation and the Chinese Pulp Corporation have collaborated in the forestation of 2,000 hectares in Vietnam, with Yuen Chi Paper Corporation investing in 150,000 hectares of forestation in Indonesia. Domestically, Yuen Foong Yu Paper Corporation has forested 1,000 hectares, with an additional 12,000 hectares by the Chinese Pulp Corporation. Finally, the paper processing industry actively engages in forestation protection efforts to encourage the efficient recycling of resources. For instance, the paper industry in advanced countries such as Finland has significantly expanded wooded areas through artificial forestation or growth of selective trees. Advanced forestation efforts will ultimately absorb greenhouse gases that pose a potential environmental threat. Therefore, the local paper industry closely adheres to governmental efforts to preserve the environment through sustainable manufacturing practices.

範例 iv.

Laboratory animals play an essential role in validating the results of medical, life science and biotechnology-related research. Realizing the importance that animals play in ensuring the quality and outcome of research, the local sector that provides laboratory animals for experimentation comprises a multidisciplinary group of highly conscientious individuals and institutions concerned with the treatment of laboratory animals in research, teaching and testing. Among the professionals involved in this market sector include veterinarians, physicians, researchers, administrators and technicians. While emphasizing the humane treatment of animals to produce beneficial research results to improve human living standards, such practitioners not only strictly comply with globally accepted laboratory and ethical standards, but also establish standards themselves with respect to use of animals in education, training and advancement of knowledge expertise. Such professional endeavors lead to advances in field expertise through continuing education programs, curriculum development and advanced research in laboratory animal science. The laboratory animal sector has established three clear objectives for further development. First, regionalized marketing and distribution must be established in supplying experimental animals to reduce laboratory costs. Next, to secure a steady supply of laboratory animals, breeding equipment and outsource vendors that provide such services must be streamlined to enhance the diversification of products and services. Finally, contacts must be established with overseas research institutes to collaboratively develop technology products in this area.

範例 v.

Widely characterized as having low investment risks yet yielding potentially high profits, the cosmetics sector has grown rapidly in the recent decade. According to the Industrial Technology Research Institute, the global cosmetics market generated revenues of $US 173,000,000,000 in 2001, with that figure expected to rise to $US 220,000,000,000 by 2006. Although most cosmetics manufacturers in Taiwan are OEM-oriented, locally produced brands are beginning to slowly gain consumer appeal. However, the lack of both qualified technology personnel in this field with managerial experience and product brand awareness poses a major obstacle to further sector growth. Previously, other than providing technical training for technical staff, local companies seldom expanded into product technology development and did not invest sufficient funds into research and development. Given the lack of technology expertise, local cosmetics companies largely imitated overseas brand products. Imported products continue to dominate the domestic cosmetics market, with product retail prices averaging twice or three times higher than local brands. Additionally, local companies generally lack the incentive to enhance the island's competitiveness level in this area, preferring to imitate overseas brands or use substandard material components in production. Alternatively, biotechnology offers viable material components and manufacturing procedures that could boost local competitiveness, such as use of natural raw extracts or advanced fermentation procedures. For instance, raw extracts or herbal additives can reduce the toxicity level in chemicals that can harm humans. With the increasing public awareness of preventive medicine, a growing health consciousness and emphasis on a youthful appearance have contributed to increased sales in anti-aging and skin whitening products. For instance, Taiyen Corporation adopts collagen in its Lu-Miel product brand, thus distinguishing itself as the first cosmetics product made in Taiwan of this nature. In competition with Taiyen for the local market, Formosa Corporation and Chang Gung University have collaboratively developed a skin maintenance product known as Forte. To foster local capabilities in the cosmetics market sector, domestic manufacturers are pursuing the following strategies: a) incorporate new raw material extracts in cosmetics products; b) continuously adopt advanced manufacturing technologies, e.g., nano technology, to upgrade the quality of cosmetics; c) develop a novel type of interface that integrates technologies used in the chemical industry; d) encourage the use of natural products and raw extracts in manufacturing that would ultimately strengthen a human's natural immune system; e) provide incentives for companies to nurture their own research capabilities for product development; f) remain attune to the latest consumer trends and preferences in the constantly fluctuating cosmetics culture; and g) encourage overseas exchanges for joint venture and technology transfer opportunities.

【參考文獻】

柯泰德（1994）。《做好英文口語簡報》。新竹：工業技術研究院。

Unit Nine

Identifying Common Errors in Professional Writing

管理英文寫作上之常見問題

簡 介

本單元包括以下三個重點：精確寫作、明白寫作及分析中英文句子結構之異同。以期解決管理英文寫作上常見之問題。

中國人從事科技英文寫作時，文詞表現自然有別於以英文為母語者。當然，文章能獲得國際人士的認可，這才是寫作的主要目的。然而，過去在台灣講授科技英文寫作方法，卻往往不切實際地要求中國人所寫的英文文章也能達到英文母語人士的同一水準。此點並不意指國人無法寫出優良的科技英文作品。但是，對於首次嘗試發表英文論文的學者或管理師而言，期望他能寫出與英文母語人士同等品質的文章，這是緣木求魚。

儘管有許多國際期刊上的論著可提供寫作者的詞句參考，但是傳統的教學方式，通常過於強調先比較別人的文句再加以應用。因此，為了刻意地表現出母語式英文的風格，過度依賴並套用期刊著作上的句子或結構，以作為著筆資源的寫作方式，可能導致肆無忌憚的抄襲，甚至不幸陷入剽竊別人作品之嫌。

另一問題是，模仿國外期刊論著的作者，也時常無法判斷自己為何在這些特定情況下使用了某些句子。他們的回答大抵上是：「因為我曾在某某期刊上的論文見過」。此種不當的英文著作方式，主要是因為國內大學或研究機構很少開授適當的科技論文寫作課程。再者，提供國內學生適當的科技論文寫作途徑，理當站在該國文化的觀點，而非著眼於以英文為母語者。國內管理師在構句時直接地習慣寫出中文的文意，再生硬地翻譯成英文句子。如此一來，使用中文慣用的語法結構硬湊成英文，常會扭曲文章內容含意的表達。因此，將主要強調的主題放於句子之前的建議，事實上是筆者考慮且瞭解國內作者的文化背景之後而提出的。再者，當管理師寫作時，希望本句與前一句子或前一段的文意相接連時，則可利用連接詞用語。此時，則須將有關前句的資訊放在句首。因此，只要在「將重要主題置於句首」及「連結效果用語」兩者之間，取得使用上的平衡，應有助於國內管理師期望以直接而通順的英文來表達文意。

但更進一步來說，管理師無法達成英文精確寫作及明白寫作二大要求。要把科技英文寫得精確並不是件容易的事情。通常管理師在文稿完成後或投寄發表前，都要不斷反覆修飾原稿避免辭不達意的窘境，由於這過程是如此的繁複，最後甚至需請專業人士代為編修始能清楚闡述作者原意。而明白寫作原則可使科技英文作者想表達的意思不會被誤解。就是要把文章中的曖昧不清全部去除，使讀者很容易的理解作品的精髓。本單元擬就這二部分逐項分類探討管理英文寫作上的問題，並於每項加以舉例說明，讓管理師可以比較一目了然。單元最後並就九個中英文句子結構之異同加以分析，同時舉例說明。

A.精確寫作（Write for Conciseness）

忙碌的專家們沒有時間去逐字的閱讀冗長的報告文件，因此很多科技管理師的文稿被拒絕或延遲發表，就是因爲他們無法精短的表達文章意念。以下有五點精確寫作的重點：

1.「常用主動語氣」（Use active voice frequently.）

・Original（原句）：

The Taguchi method is assessed in this study so that the effect of various factors on performance variation can be understood.

・Revised（修改過語句）：

This study assesses the Taguchi method to understand how various factors affect its performance variation.

・Original（原句）：

A description of the layout cost evaluation method is made by introducing the following notations in this section.

・Revised（修改過語句）：

This section introduces the following notations to describe the layout cost evaluation method.

2.「動詞替代名詞」（Use verbs instead of nouns.）

・Original（原句）：

Not only is the variability of the control factors considered by the proposed procedure, but implementation of related tasks is performed by that same procedure.

・Revised（修改過語句）：

In addition to considering the variability of the control factors, the proposed procedure implements related tasks.

・Original（原句）：

Solution of environmental problems is heavily dependent on responsiveness of the local community to governmental legislation.

・Revised（修改過語句）：

Solving environmental problems heavily depends on how the local community responds to governmental legislation.

3.「強有力的動詞」（Create strong verbs.）

・Original（原句）：

A stipulation of the agreement is that the parties are in compliance with the rules.

・Revised（修改過語句）：

The agreement stipulates that the parties comply with the rules.

・Original（原句）：

The guideline makes it specific that all parties will be notified by the proper authorities when a problem arises.

・Revised（修改過語句）：

The guideline specifies that the proper authorities will notify all parties when a problem arises.

4.「避免過度使用It 及There開頭句」（Avoid overusing sentences that begin with It and There.）

・Original（原句）：

There is a need for limitation of the number of users by the systems manager.

・Revised（修改過語句）：

The systems manager must limit the number of users.

・Original（原句）：

It is necessary that negotiations on the final conditions be undertaken by both sides immediately.

・Revised（修改過語句）：

Both sides must negotiate the final conditions immediately.

5.「除去重複及不必要的措詞」（Delete redundant and needless phrases.）

・Original（原句）：

It is well known that a majority of Taguchi method applications have the capacity for addressing a single-response problem.

・Revised（修改過語句）：

Most Taguchi method applications can address a single-response problem.

・Original（原句）：

It is our opinion that Windows 2003 is for all intents and purposes a good operating system in a situation in which the user has many requirements.

・Revised（修改過語句）：

We believe that Windows 2003 is a good operating system when the user has many requirements.

B.明白寫作（Write for Clarity）

明白寫作延伸精確寫作指導原則，更進一步把重點放在如何讓作者想表達的意思更明顯。以下有六點明白寫作的重點：

1.「主詞及動詞必須前後呼應：如果主詞的單複數與動詞不能配合，不僅讀者感到困惑，同時句子的邏輯也會發生問題。」（Ensure subject and verb agreement.）

・Original（原句）：

Neither the processing parameters nor the amount of polymer melt injection solely determine a successful gas channel design.

・Revised（修改過語句）：

Neither the processing parameters nor the amount of polymer melt injection solely determines a successful gas channel design.

・Original（原句）：

The delay time along with gas injection points are important in producing injection-molded parts of quality.

・Revised（修改過語句）：

The delay time along with gas injection points is important in producing injection-molded parts of quality.

2.「代名詞必須清楚的使用：如果代名詞所指的人物或事物不能交待清楚，也是徒增讀者困惑。」（Ensure that pronoun references are clear in meaning.）

・Original（原句）：

The machines can process the family of parts, ensuring that it reaches full capacity.

・Revised（修改過語句）：

Capable of processing the family of parts, the machines reach full capacity.

・Original（原句）：

The department chairman told the research assistant that she needed to perform the simulations if the experiment fails.

・Revised（修改過語句）：

The department chairman told the research assistant to perform the simulations if the experiment fails.

3.「句子的結構須有一致性：科技英文寫作中，句子的建構必須有一致性。」（Create sentences parallel in structure and meaning.）

・Original（原句）：

The two-step procedure not only identifies those factors that significantly affect the signal-to-noise (SN) ratio, but also the levels that maximize SN are found.

・Revised（修改過語句）：

The two-step procedure not only identifies those factors that significantly affect the signal-to-noise (SN) ratio, but also finds the levels that maximize SN.

· Original（原句）：

The Taguchi approach provides a combination of experimental design techniques with quality loss considerations and that the average quadratic loss is minimized.

· Revised（修改過語句）：

The Taguchi combines experimental design techniques with quality loss considerations and minimizes the average quadratic loss.

4.「去除修飾語所造成的問題：修飾語必須放在所要修飾的字之旁。」（Eliminate modifier problems.）

· Original（原句）：

Having a large database, much information is contained in the system's network.

· Revised（修改過語句）：

Having a large database, the system's network contains much information.

· Original（原句）：

The container held by the experimenter with many leaks must be replaced.

· Revised（修改過語句）：

Held by the experimenter, the container with many leaks must be replaced.

5.「再次檢查錯誤的比較詞及粗心的疏漏：不合邏輯及不完整的比較詞造成更多的含糊不清，同時注意不小心漏掉的字。」（Double check for faulty comparisons and omissions.）

· Original（原句）：

Our algorithm is more accurate with respect to computational time.

・Revised（修改過語句）：

Our algorithm is more accurate than conventional ones with respect to computational time.

・Original（原句）：

The trade school's dropout rate is lower than the university.

・Revised（修改過語句）：

The trade school's dropout rate is lower than that of the university.

6.「避完不必要的句中轉換：作者應避免句中不必要的主詞、時態及語態之轉換。」（Avoid unnecessary shifts in a sentence.）

・Original（原句）：

Give consideration to the external factors and the surrounding environment must be assessed.

・Revised（修改過語句）：

Give consideration to the external factors and assess the surrounding environment.

・Original（原句）：

Not only can the devices be used to copy CDs in case of damage to the original, but it can also be used for backing up hard drives.

・Revised（修改過語句）：

In addition to copying CDs in case of damage to the original, the devises can also back up hard drives.

C. 分析九個中英文句子結構之異同（Avoid Chinese-English Colloquial Habits in Writing）

對從事科技英文寫作的管理師而言，首先要面對的工作便是熟知他自己在寫作表達上的習慣，例如在寫作時直接翻譯中文或過度依賴其他類似文章上的句子，這些句子的意思除了原作者想要表達的以外，通常常有其他的引申義。不論如何，過度使用一些連作者自己都不熟知的寫作習性，勢必使得科技文章本身的內容變得含糊不清，且在意義表達上造成混淆。雖然每個作者都有他自己的需求，而大部分的寫作模式又都含混不分；但是，不論是對一位寫作科技文章的新手或對極有經驗的管理師而言，上述國人易犯之習性總使得國內管理師所寫的科技論文具有某些共通的缺點。管理師可先從回顧自己以前寫的英文開始，確認哪些詞句已爲國人科技論文作者共通特有的寫作習性，然後，以下列九點所建議的寫作方法，再來看看文章的內容是否面目一新，更易瞭解。

1.「著手寫作論文時，切勿過度使用一些課本中常見而生硬的詞句，盡量使文章中的英文直接而流暢。」（Prevent overuse of traditional textbook phrases.）

・Original（原句）：

It may be said that computers have the ability to incorporate kinds of equipment in order that the user is in a position to interact with the computer.

・Revised（修改過語句）：

Computers can incorporate equipment so that the user can interact with the computer.

・Original（原句）：

There is a chance that the college will install 100Mbps cards notwithstanding the fact that they have served the school well in the past history.

・Revised（修改過語句）：

The college might install 100Mbps cards although they have served the school well in the past.

2.「把最重要的主題、子句等放在句首，以求凸顯你主要的意思或重點，而且能容易地為讀者所領會。」(Make your intended meaning more visually accessible.)

・Original（原句）：

There is no need for the assessment of the production capacity of the factory floor to be undertaken by the foreman.

・Revised（修改過語句）：

The foreman does not need to assess the production capacity of the factory floor.

3.「英文已經規定了完整的過去／現在／未來等動詞時態，但是，國內的科技論文寫作者仍然習慣地把一些不必要的時間介系詞置於英文句首。時間介系詞片語常被用來作為開頭的字有“during”、“until”、“as”、“before”、“after”、“when”、“since”及“whenever”等。但是，當用來表示事件的順序或當作不同步聚的轉接詞時，時間介系詞的使用就變得適宜而需要了。」（Avoid overemphasis of clauses involving time.）

・Original（原句）：

Currently, there are more than fifty varieties of bulk pharmaceuticals produced in Taiwan.

・Revised（修改過語句）：

More than fifty varieties of bulk pharmaceuticals are produced in Taiwan.

4.「保持如「名詞＋動詞＋受詞／（介系詞）」的簡單而直接的句型，並且避免把比較性介系詞放在句首。」（Avoid overemphasis of clauses involving comparison.）

・Original（原句）：

Compared to conventional approaches, the proposed scheme spends less computational time.

・Revised（修改過語句）：

The proposed scheme spends less computational time than conventional approaches do.

5.「使用如「名詞＋動詞＋受詞／介系詞」般的直接簡單句。除非：(1)介系詞在如下的句型中（名詞＋動詞＋名詞＋動詞）描述動作；(2)介系詞用作科技條件的限制語；(3)介系詞作為連接二個句子或段落的轉接語，否則介系詞不擺於句前。」（Avoid overemphasis of "domains" which tend to push the author's intended meaning towards the end of the sentence.）

・Original（原句）：

For determination of the optimum conditions for the nominal-the-best robust design problems, this work is conducted to develop a simple procedure to do so.

・Revised（修改過語句）：

This work develops a simple procedure to determine the optimum conditions for the nominal-the-best robust design problems.

6.「使用如下的直接簡單句「名詞＋動詞＋受詞／介系詞」。但是，表示連接性質的介系詞（副詞或其他相關的片語）在連接二個句子或二個段落時，則反而常用且有用。」（Use transitional phrases to link sentences and promote the manuscript's flow.）

・Examples:

Under those circumstances, For such a reason, In such a case,
In light of above developments, As mentioned earlier, According to those results,
Above results suggest that... If the above condition is satisfied,

7.「藉著以下的方法達到置換動詞的目標：(1)在不定動詞和動名詞間取得平衡，避免過度使用不定動詞；(2)使用動詞的名詞形式；(3)找出動詞的同義字，利用同義字的替換，避免在同一篇文章中一直使用同一個動詞。」（Provide variety in verb form and selection.）

・Original（原句）：

Our results fit very good with those of Wu et al.

・Revised（修改過語句）：

Our results (correlate well with OR closely correspond to) those of Wu et al.

8.「一個句子只表達一或兩個主要意思，可以避免句子太長的毛病。以下三個方法可使繁長句子變得簡單易懂：(1)將一句分為二句，意即第一子句（結果）＋第二子句（位置、條件、目的、結果表示等特定的範圍）；(2)在句子中間使用分號以區分句子中的兩個重點意思，使其更清楚；(3)盡量在一個句子中只表達一個意思，最多不超過兩個。」（Avoid excessively long sentences.）

・Original（原句）：

Since the climate of Orchid Island is relatively hot and humid compared to Taiwan, the research team from the research institute decided to return them to their native habitat on Orchid Island to allow the butterflies to be successfully born.

・Revised（修改過語句）：

The climate of Orchid Island is hotter and more humid than Taiwan, explaining why the team from the research institute decided to return the butterflies to their native habitat on the island for their successful birth.

9.「在科技文章中，一般而言使用第二人稱較用第一人稱為佳，其理由為：(1) 研究內容理當客觀的，過度使用第一人稱會讓讀者有內容主觀及具有成見感覺；(2) 通常有將最重要和強調的部分擺在句首的傾向。」（Avoid overuse of first person.）

・Original（原句）：

From the above literature having been mentioned, we can find that...

・Revised（修改過語句）：

Above literature suggests that...

【參考文獻】

柯泰德（1993年）。《精通科技論文（報告）寫作之捷徑》。新竹：清蔚科技。
柯泰德（2000年）。《科技英文編修訓練手冊》。新竹：清蔚科技。

Unit Ten

Writing Persuasive Employment Application Letters

求職申請信函

簡 介

本單元描述如何撰寫求職申請信函，包括：經由廣告、朋友及其他相關訊息得知的工作申請，對未來工作的個人學術及相關專業特質總結，讚揚要申請工作的機構及說明其對個人工作目標所產生的利益，總結主要論點及邀請讀者給予面試機會。

A.經由廣告、朋友及其他相關訊息得知的工作申請

範例 i.

In response to your recent advertisement in the September edition of Quality Management Magazine for an executive coordinator of marketing projects in your healthcare products company, I feel that my strong academic background and practical experiences make me highly qualified for this position.

範例 ii.

The position of a pharmaceutical sales representative that you recently advertised in the February 5, 2007 edition of The China Times closely matches my career direction and previous academic training.

範例 iii.

I would like to apply for the marketing position in the information technology sector that you advertised recently in the March issue of Information Today. Your company would provide me with an excellent environment not only to realize fully my career aspirations, but also to apply theoretical knowledge management concepts taught in graduate school in a practical work setting.

範例 iv.

From a recent March 15th posting on the Nursing Today website, I learned that your hospital is recruiting for a supervisor in the emergency care department. Fully understanding that your hospital hopes to operate an emergency care department that can coordinate its efforts with an emergency medical network in Taipei, I am confident of my ability to handle required administrative tasks, especially in light of my solid nursing experience and management skills.

範例 v.

I would like to apply for the managerial position posted recently on your healthcare organization's website. Abundant experience of administrative work while in the military, and extracurricular activities in the student association at university, will definitely help me become oriented with my new responsibilities at your company. My familiarity with several analytic methods in decision science, gained during graduate school, is another strong asset that I bring to your company. I can apply such methods more flexibly after securing employment in a non-profit organization such as yours. Moreover, my previous academic and work experience will enable me to explain results of analyses more clearly to enable the organization to reach management decisions efficiently.

B.對未來工作的個人學術及相關專業特質總結

範例 i.

I received a Bachelor's degree in Industrial Management from National Cheng Kung University, where the departmental curriculum sparked my interest in various directions. I became especially interested in strategic management, particularly for the hi-tech sector. Several years of academic study have left me with the deep impression that study is not just for securing employment. As indicated in the attached resume, my recent completion of a Master's degree in Business Administration Imagery attests to my commitment to pursuing a marketing career. While attempting to understand the recent expansion of the local cosmetics sector, I focused my graduate study on the role of customers relationship management in complying with consumer demand. The results of my research have already contributed to efforts of financial planners to estimate long term trends and to devise appropriate marketing strategies in response. Specifically, I am interested in how various statistical methods can accurately forecast future growth trends in this burgeoning sector - an area of research that has received increasing attention in recent years. Working in your organization would hopefully allow me not only to pursue some of my above research interests, but also to contribute to the further development of this thriving sector.

範例 ii.

Commercial development planning has enthralled me since I took part in a business management training course sponsored by the Council of Labor Affairs. I also focused on development planning in the medical sector while studying in the Department of Healthcare Management at Yuanpei University and later for a Masters in Business Management from the same university. As you can see in the attached resume, while pursuing a graduate degree, I learned of advanced theories in my field and acquired practical training to enhance my ability to identify and resolve problems efficiently. Moreover, closely studying business practices in the medical sector at undergraduate and graduate school has equipped me with the competence to contribute to the development of management strategies to resolve unforeseeable problems efficiently, hopefully at your company.

範例 iii.

Besides my strong interest in information technology, I have acquired many valuable research experiences through graduate studies in Business Management at National Taiwan Ocean University. These two ingredients are definitely crucial to my fully realizing my career aspirations. I will be able further to refine my skills if employed at your company. My logical competence and analytical skills reached new heights during two years of intensive graduate level training. As detailed in the attached resume, related project experiences have greatly strengthened my independent research capabilities and statistical as well as analytical skills. Your company offers a competitive work environment for highly skilled professionals: these ingredients are essential to my continually improving my knowledge, skills and expertise in the above area.

範例 iv.

From my early nursing experience to my current work as a professional nurse in an emergency care department, I have been involved in many efforts that require critical patient care. They have included responding to the devastating Chi Chi earthquake that hit Taiwan on September 21, 2000, setting up a medical station at San Xia Da Bao River, and counteracting the deadly SARS virus in our hospital environment. While my job often requires me to respond to many unforeseeable circumstances, I remain fascinated by the dynamics of my profession, and enjoy a deep sense of satisfaction when helping others. My recent completion of a master's degree in Business Management reinforces my commitment to quality excellence in this profession.

範例 v.

As indicated in the attached resume, my graduate degree in Business Administration at National Central University offered specialized curricula that not only strengthened my knowledge of modern business practices and English writing skills to publish my research findings, but also equipped me with the necessary skills to contribute significantly to the workplace. My diverse academic interests and strong curricular training reflect my ability not only to see beyond the conventional limits of a discipline and fully comprehend how the field relates to other fields, but also to apply strong analytical and problem-solving skills.

C.讚揚要申請工作的機構及說明其對個人工作目標所產生的利益

範例 i.

Your hospital offers comprehensive and challenging training for marketing professionals eager to continuously upgrade their competence. Such training could provide me a marvelous opportunity to put my above knowledge and expertise into practice. If employed in your company, I will apply my professional knowledge of how to understand fluctuating consumer preferences given the increasingly shorter product life of healthcare goods.

範例 ii.

I am even more confident that I can significantly contribute to your franchise. While offering famous medical product brands from the United States in the Taiwan market, your franchise prioritizes quality and the work professionals in a diverse range of medical fields. Specifically, I hope to contribute to your marketing efforts, logistics management and medical research in areas that are likely to grow in the future. As a community pharmacy chain, your franchiser has distinguished itself in overcoming operational difficulties and maintaining good discipline to manage its business units effectively.

範例 iii.

As a leader in the information technology sector, your company is renowned for its state-of-the-art products and services, as well as outstanding product research and technical capabilities. As I hope to become a member of your corporate family, the particular expertise developed in graduate school and my strong academic and practical knowledge skills will definitely make me an asset to any collaborative product development effort. Offering more than just my technical expertise, I am especially interested in how your company’s marketing and related management departments reach strategic decisions. Employment at your company will undoubtedly expose me to new fields as long as I remain open and do not restrict myself to the range of my previous academic training.

範例 iv.

Renowned for its efforts to remain abreast of governmental strategies and current trends, your hospital will establish a second medical center in Taipei County in the near future. I am most impressed that this new center will provide high-quality medical treatment to more critical patients in a relatively short time. With my decade of experience in emergency care departments and graduate degree in Business Management, I believe that I can significantly contribute to efforts to train emergency care professionals and assess the process and preparation of emergency medical staff.

範例 v.

As a renowned leader in the long-term healthcare sector, your company has impressive organizational objectives, combined with a strong management structure and diversity of training courses. This strong organizational commitment is reflected in the high quality services that you provide to your customers. As a member of your corporation, I hope to participate actively in your company's external affairs. My work experience and solid academic background will enable me to comprehend and familiarize myself with all of the commercial practices of your healthcare organization in a relatively short time.

D.總結主要論點

範例 i.

My professional knowledge will facilitate your efforts to expand upon your already solid market position. I am confident that my solid academic background in management will prove to be an invaluable asset to your organization.

範例 ii.

I firmly believe that if I successful in securing employment at your company, my strong academic and practical knowledge, curricular and otherwise, will enable me to contribute positively to your corporation. In summary, my marketing research and excellent analytical capabilities will support your company's commitment to offering quality products and services.

範例 iii.

In sum, my ability to excel in information technology proved especially effective in devising marketing strategies for research purposes. Despite lacking knowledge of a particular topic at the outset, I quickly absorb new information and adapt to new situations.

範例 iv.

Given my experience, I will definitely be able to contribute to your hospital's efforts to satisfy the demands of medical consumers while effectively controlling overhead. In addition to my background in the nursing profession, I strive constantly to integrate my substantial management experiences with theoretical knowledge, making me more adept in making the correct strategic decisions in our hospital's emergency medical unit.

範例 v.

In addition to a solid academic background, a good manager should have strong communicative, organizational and management skills. I am confident that I possess them.

E.邀請讀者給予面試機會

範例 i.

I look forward to meeting with you in person to discuss this position further.

範例 ii.

I will contact you shortly by telephone with the hope of scheduling an interview.

範例 iii.

I believe that your company will find this a highly desired quality. Please contact me at your earliest convenience to schedule an interview.

範例 iv.

I am confident of my ability to contribute significantly to your efforts to elevate the quality of healthcare management if given the opportunity to work in your organization. The attached resume includes my contact details. Please give me this opportunity.

範例 v.

I look forward to meeting with you in person to discuss in detail what this position encompasses.

◎Appendix Sample Employment Application Letters

Dear Mr. Su,

經由廣告，朋友及其他相關訊息得知的工作申請 Regarding the information management position advertised recently in the December issue of Information

Today, I am a competent candidate. My project management experience has enabled me to deal carefully with others and resolve disputes efficiently. My love of challenges will enable me to satisfy constantly fluctuating customer requirements in information integration projects, hopefully at your company. 對未來工作的個人學術及相關專業特質總結 Having devoted myself to developing information systems in the semiconductor industry for over a decade, I have developed a particular interest in enhancing work productivity using the latest information technologies. I have also spent considerable time in researching system integration for manufacturing applications on UNIX-based systems. As shown in the accompanying resume, critical thinking skills developed during undergraduate and graduate training have enabled me not only to explore beyond the surface of manufacturing-related issues and delve into their underlying implications, but also to conceptualize problems in different ways. While participating in several MOSEL group projects, I also learned how to address supply chain-related issues to broaden my perspective of potential applications of finance and decision making; these areas are now my main focus.

讚揚要申請工作的機構及說明其對個人工作目標所產生的利益Renowned for effectively dealing with unforeseeable emergencies and enhancing customer services, your company has established a vision that deeply impresses me. Moreover, I am attracted to your company's advanced financial information system for analyzing business transactions models, a system which will equip me with the competence to contribute more significantly to your organization's excellence in marketing. If I am successful in gaining employment in your company, both my solid academic training and my research on information system development will make me a strong asset in your efforts to upgrade e-business operations, such as online queries, payments and account transferals.

總結主要論點 I am confident that my work experience in software development has equipped me with the necessary competence to address effectively information technology-related problems in your company. 邀請讀者給予面試機會 Please contact me at your earliest convenience to schedule an interview.

Yours truly,

Christie Wang

Dear Dr. Thomas,

Having graduated recently from the Master's degree program in Business Management from Yuanpei University in Taiwan, I focused on finance-related research and often drew upon related concepts from your publications and often cited them while writing my master's thesis. I learned recently that you are seeking research assistants for a special research project. Such an opportunity to work under your supervision would allow me to become more proficient in this line of research. My academic advisor often praises your research in class, commenting on your leading role in finance research.

Besides its immense popularity given the need for efficient organization, finance is constantly evolving, especially in developing countries such as China and India. Countries such as yours offer a valuable resource for finance students in developing countries to increase not only their research capabilities in finance by understanding how financial markets worldwide affect each other, but also their English language skills. Financial experts must adopt a global perspective to remain abreast of current trends.

Despite my academic background, I lack the necessary professional skills that are required to achieve the above goals. This explains why I am excited about research opportunities such as yours to compensate for the deficiencies in my research approach. Nevertheless, knowledge acquired in graduate school has instilled in me a solid foundation for professional experiences such as the opportunity to work in your laboratory. The attached resume describes my professional experiences in the above area. I look forward to hearing from you soon as to whether such cooperation is possible. Thanks in advance for your kind assistance.

Sincerely yours,

Danny Wu

Unit Eleven

Writing Concise Professional Training Application Letters

專業訓練申請信函

簡 介

本單元描述如何撰寫專業訓練申請信函，包括：專業訓練申請、某人學歷及工作經驗概述、讚揚提供專業訓練的機構、專業訓練細節解釋及結語。

A.專業訓練申請

範例 i.

To build upon my academic and professional experiences, I would like to serve as a self-supported guest researcher in the oncology department of your hospital for a three-month period, hopefully during my upcoming summer vacation.

範例 ii.

My academic advisor, Dr. Cheng, recommended that I contact you regarding the possibility of a guest researcher stay in your laboratory, hopefully for a six-month period under your direct supervision. Despite my strong academic background and numerous work experiences, I hope to gain further training at your hospital owing to its commitment to excellent medical image processing as well as advanced PACS instrumentation. In addition to improving my technical expertise, I am also interested in enhancing my management proficiency within your organization.

範例 iii.

Eager to strengthen my expertise in optimizing the results obtained from the picture archiving communication system (PACS) at our hospital's Medical Imagery Department, I hope to serve in your laboratory as a self-supported guest worker for six months to compensate for my lack of training in this area. As a leader in the medical field, your hospital would provide me with me with many collaborative opportunities not only to provide better medical care for patients through your solid training, but also to improve my expertise continually.

範例 iv.

Eager to understand your dynamic work, I would like to arrange for a three-month stay as a visiting medical physicist in your cancer research center for tumor growth.

範例 v.

The opportunity to serve in a self-supported guest researcher position in your radiation dosimetry laboratory would help me to become more proficient in this line of research. As a highly adept investigator in the laboratory, I have learned not only how to comprehend fully how seemingly opposite fields are related to each other, but also to handle complex projects that forced me to apply theoretical concepts in a practical context. Such practical work experience has greatly enhanced my competence in accumulating pertinent data and analyzing problems independently.

B.某人學歷及工作經驗概述

範例 i.

The attached resume and recommendation letters provide further details of my solid academic background and professional experiences. As a graduate student of Medical Imagery at Yuanpei University, I actively participated in a project aimed at identifying prognostic factors of breast cancer and subsequently developing an effective prognostic method to increase the survival rate of breast cancer patients. The identified prognostic factors provide a valuable reference for radiologists in devising a therapeutic treatment program for patients. The prognostic factors, as identified in that research effort, can also benefit the cancer patients in your hospital, an organization that already has an excellent reputation in the medical field and has gained international recognition for its own research advances.

範例 ii.

My interest in medicine and physics, since my childhood, has led to my successful completion of both a Bachelor' s degree in Radiology Technology and a Master' s degree in Medical Imagery from Yuanpei University. Since then, I have been working as a radiology technician in the radiology department of a hospital, which combines digital technologies and the Internet. Undergraduate and graduate level courses in radiochemistry in the medical imagery field instilled in me a wide array of theoretical concepts related to radiopharmaceutical synthesis. My graduate school research often involved deriving complex radiopharmaceutical synthesis models and modifying radiopharmaceutical practices in nuclear medicine to meet my research needs. I have also attended several international conferences on radiology technology, further widening my exposure to the radiochemistry profession.

範例 iii.

As I recently completed my Master' s degree in Medical Imagery, graduate school has oriented me on how to integrate diagnostic programs with the Internet and related technologies, greatly improving a patient' s outcome. Graduate school also enabled me not only to grasp the clinical implications of different diagnostic tests, but also to operate medical instrumentation proficiently. These skills help me to ensure that a patient receives an accurate diagnosis.

範例 iv.

As you can see in the attached resume, my undergraduate and graduate training in radiotechnology and medical imagery heavily stressed close collaboration among researchers. I became adept in applying various radiation detection methods. As my graduate research focused on detecting contamination during clinical practice, I had to familiarize myself with the underlying causes of contamination, the extent of the injury caused to humans and a wide array of preventive methods. These academic and professional experiences increased my confidence in planning radiological treatments, and implementing radiation detection and protection strategies during therapeutic treatment.

範例 v.

Having received a Bachelor's degree in Atomic Science and a a Master's degree in Medical Imagery, I am well aware of the theoretical and practical issues around radiation. As reflected by the attached resume, my undergraduate studies in atomic physics led me to thoroughly understand how radiation affects materials. Additionally, my active participation in radiation dosimetry-related projects taught me how to measure an actual dose from any radiation source, including photons, electrons, neutrons, X-rays and gamma rays. Hopefully, you will find these skills to be invaluable to any research collaboration in which I am involved at your laboratory.

C.讚揚提供專業訓練的機構

範例 i.

The variety of research projects and departments within your company committed to implementing them is quite impressive, explaining why you have taken a leading role in the medicine and pharmaceutical fields. Working at your company would definitely promote my professional development. Following my grasp of medical imagery expertise in graduate school, I believe that my solid academic training and practical knowledge will contribute to your company's efforts to elevate its reputation and new technology capabilities, even during the short three-month period at your laboratory.

範例 ii.

Through this training opportunity in your laboratory, your highly skilled professionals would provide me with an excellent research environment, advanced equipment and related resources to enhance my research capabilities so that I can thrive in this dynamic profession. With a long tradition of commitment, your hospital offers extensive training courses for technical staff in all hospital departments to maintain competitiveness. Working in your organization, even for a short time, would definitely benefit my professional development.

範例 iii.

As evidenced by your highly respected training courses on nuclear medicine, your hospital possesses state-of-the-art equipment and expertise in handling stroke patients. For instance, your excellent staff has perfected the skill of easily distinguishing ischemia from hemorrhaging. If granted this research opportunity, I would bring to your organization a solid academic background and practical expertise that can hopefully contribute to your ongoing efforts. Moreover, having passed an extremely difficult entrance examination for medical professionals in this field, I believe that my expertise of imagery medicine will be an asset to any clinical department to which I belong.

範例 iv.

Your cancer research center has distinguished itself in developing novel radiotherapy methods that have greatly benefited patients, as evidenced by your frequent publications in international journals. Your center is widely admired for its strict adherence to quality standards in tumor therapy, as well as its frequent journal publications on novel radiotherapeutic procedures and clinical findings. If successful in securing this visiting researcher position, I will bring my professional knowledge in such areas as detecting radiation dose in the workplace, planning appropriate radiotherapy and devising appropriate shielding for a radiotherapy room.

範例 v.

As a leader in radiation dosimetry research, your laboratory would provide an excellent environment for me to build on my above academic experiences so that I can expand my research activities and grasp many helpful concepts related to the latest technological trends. The impressive training courses that your laboratory offers reflect your excellence in leadership and commitment to staff excellence.

D.專業訓練細節解釋

範例 i.

As for the details of this training practicum, identifying adequate therapeutic treatment and prognostic factors is essential in radiology technology - a field in whose managerial aspects I am very interested. As for my professional interests, I have always been interested in identifying prognostic factors of breast cancer or, more specifically, those factors can that can elevate the curative rate for patients during treatment. In this area of research, the potential technical and medical sector opportunities appear to be limitless.

範例 ii.

As for details of this guest internship, your department offers state-of-the-art instrumentation and clinical training of those involved in researching PET/CT-related topics.

範例 iii.

During this training opportunity, I will be especially interested in how medical images facilitate the diagnosis and treatment of diseases. In particular, computers with valuable medical software can provide clinical physicians with data that can help determine the course of medical care. Training at your hospital would enable me to create precise anatomic images to confirm a specific malady.

範例 iv.

Regarding my specific interests during this research stay, I am especially interested in the importance of dose detection to radiation security. This interest demands not only becoming proficient in the use of many radiation detection methods, but also understanding the applicability of such methods in a clinical hospital setting. I believe that I possess the necessary practical and theoretical skills as a medical physiologist to contribute to a patient' s well-being and simultaneously maintain radiation security.

範例 v.

As for the details of this proposed researcher position, I am intrigued with the increasing importance of radiotherapy, especially given the rising cancer death rate. Radiotherapy is especially attractive since it does not involve injecting the patient, yet yields curative effects rapidly.

E.結語

範例 i.

Please let me know if you require additional materials. I look forward to your favorable reply.

範例 ii.

Such exposure would definitely further my knowledge expertise. I look forward to your thoughts regarding this proposed residency.

範例 iii.

Let me know if you require materials in addition to the enclosed resume and recommendation letters. I look forward to your favorable reply.

範例 iv.

I believe that your organization would be an excellent starting point for me to begin on this career path. I look forward to your thoughts regarding this proposed stay.

範例 v.

Hopefully, by working directly under your supervision, I will gain further exposure to the latest techniques in this field. Please do not hesitate to contact me for an interview if this proposal is feasible.

◎ Appendix Sample Professional Training Application Letters / 專業訓練申請信函

Dear Dr. Li,

專業訓練申請I would like to arrange for a stay as a self-supported guest worker in your biotechnology company for six months. The fundamental and advanced research capabilities I acquired in graduate school have not only nurtured my talent in approaching biotechnology through a multidisciplinary approach, but also widened my range of interests and helped me to grasp fully the latest concepts in biotechnology.

某人學歷及工作經驗概述Having immersed myself in the field of radiation for quite some time, I recently completed a Master' s degree in Life Sciences at

National Cheng Kung University, with a particular interest in researching biology-related topics. While pursuing this Master' s degree, I conducted biotechnology-related research at the Animal Technology Institute of Taiwan. As shown in the attached resume, graduate level research prepared me for the rigorous demands of experimentation and, then, publication of experimental findings in domestic and international journals. Overall, my participation in research projects that encompassed seemingly unrelated fields reflects my willingness to absorb tremendous amounts of information and manage my time efficiently, an attribute which I believe that your company looks for in its research staff.

讚揚提供專業訓練的機構 As a leader in the biotechnology field, your company has been able to combine commercial success with innovation. Your company has also distinguished itself in the healthcare sector. I am especially impressed with your company' s creativity in using standard operating procedures to create state-of-the-art product technologies. Participating in the innovative research projects at your company would further strengthen my expertise in biotechnology and, hopefully, contribute to your ongoing efforts.

專業訓練細節解釋I am increasingly drawn to biotechnology, an emerging global field in the new century. Its emergence reflects an increasing emphasis on health, which accompanies an increase in the elderly population worldwide. To become proficient in this area, I must acquire further laboratory experience, explaining why I am seeking a valuable training opportunity at your company. The opportunity to work in a practicum internship in your company would provide me with an excellent environment not only to realize my career aspirations fully, but also to improve my own technological expertise. 結語Please contact me if such an opportunity arises.

Yours truly,

John Wang

Hello,

As a self-supported guest worker, I would like to receive specialized training in procurement management from your company. I received a bachelor's degree from Chia Nan University of Pharmacy and Science in 2002. Undergraduate curricula exposed me to hospital administration, medical stock supply and logistics management. Given my work in the Procurement Department in the Northern Region Alliance of DOH Hospital, I want to acquire more academic skills and practical exposure to become more proficient in the above areas.

Since completing our country's compulsory military service, I began working in hospital administration, including enterprise planning, marketing and quality care in hospitals. Such experience instilled in me a solid foundation in hospital operations. Later, I began working in a hospital in Keelung as a liason with the National Health Administration in procurement of medical instrumentation and materials. Through the news media, Internet, magazines and referrals from colleagues, I learned of your company's unique management style in procurement and how it streamlined its administrative procedures to achieve efficient and convenient purchases.

Given the opportunity, receiving specialized training in procurement management from your company would greatly foster my administrative skills. If granted this opportunity, I would like spend a month, self-supported stay in your company, in which I will make all necessary accommodation and transportation arrangements.
I am especially interested in working under the supervision of the procurement division of your company. I look forward to your favorable reply. Please do not hesitate to contact me if you require further details.
Sincerely yours,
Thomas Lin

Dear Sir/Madman:

I would like to apply for the research fellow position that you recently advertised in the September edition of Hospital Administrator Today. I received a Bachelor' s degree in Healthcare Management and a Master' s degree in Business Management in 2005 and 2007, respectively, from Yuanpei University. During this period, I acquired specialized knowledge and professional skills in healthcare and business management. I was encouraged to learn of your academic and professional accomplishments.

I am well aware of the high professional standards that your organization holds in instruction and research. In addition to your distinguished faculty from both locally and abroad, your organization has actively participated in several nationwide research projects under the auspices of the National Science Council and the National Institute of Health. As is well known, your organization heavily prioritizes training diligent, perseverant, frugal and trustworthy students who have strong theoretical and practical foundations. I am especially impressed with the practicum internship programs that you sponsor in conjunction with Chang Gung Memorial Hospital, the Formosa Plastics Group and other institutions with similar work-study programs.

After receiving my master' s degree in Business Management, I plan to continuously upgrade my specialized management skills. Through my work, I came to realize that your school would be an ideal environment for me. My healthcare and business management background would prove a valuable asset to any research effort I am involved in at your institution.

As a research fellow at Chang Gung University, I want to reiterate my strong interest in researching topics at your school that match my background. Please feel free to call me (tel. XXXXXXXX) if I can provide additional information.

Yours sincerely,

Jean-Yi Chen

Dear Professor Jones,

My academic advisor, Professor Hsu, suggested that I contact you regarding the possibility of receiving technical training as a visiting researcher in your laboratory. As a graduate student in the Institute of Business Management at Yuanpei University (YPU), I would like to receive training in customer service-related issues, specifically how marketing and an increasing number of customer-oriented administrative systems can facilitate customer relations. I hope to work directly under your supervision.

Having studied marketing since high school, I am pursuing a master's degree in the Institute of Business Management at YPU. Hoping to research science and technology policy issues, I am specifically concerned with how to measure income and consumption-related factors in order to more fully understand consumer behavior in China. To do so, I must first strengthen my statistical analysis capabilities, hopefully in your laboratory.

After participating in an intensive English language learning course in Australia, I hope to join your laboratory for four months this year, preferably from September to December. I hope to become familiar with the latest trends in customer service and marketing. The proposed format is that of a self-supported guest worker at your laboratory. I consider myself diligent and able to grasp new concepts easily.

As for accommodations, I would like to live in your university's dormitory if possible. Thanks in advance for your careful consideration. I look forward to our future cooperation.

Sincerely yours,

Ivy Guo

Unit Twelve

Writing Convincing Employment Recommendation Letters

求職推薦信函

簡 介

本單元描述如何撰寫求職推薦信函，包括：簡介、推薦人的資格、被推薦人跟求職有關的個人特質及信函結尾。

A.簡介

範例 i.

Recommending Mary Li for employment in your globally renowned corporation is indeed an honor.

範例 ii.

The opportunity for Jerry Su to gain professional competence in an eminent organization such as yours is a marvelous one.

範例 iii.

I have encouraged Matt Chen to seek employment with your company for quite some time.

範例 iv.

I was pleased to hear that Susan Chuang is seeking employment in your company.

範例 v.

I am pleased to recommend John Chang for employment in your organization.

B.推薦人的資格

範例 i.

Over the past two years that I have supervised her in the product development group, I have been most impressed by Mary's energy, as evidenced by her seemingly endless perseverance in handling tedious and detailed tasks. Also, her proactive approach towards learning has undoubtedly enabled her to improve her knowledge skills continuously. Given your company's commitment to excellence in product technology and customer service, I cannot think of a better qualified employee.

範例 ii.

As his academic advisor who supervised his doctoral level research and dissertation, I feel that I am in a good position to assess this highly competent individual. Jerry's unique ability to integrate and explain seemingly contradictory concepts to those outside of his field of expertise is invaluable in the workplace as it can help provide enough details to managers to enable them to make management decisions based on that information. As a leader, he applied his excellent communication skills to identify fellow classmates' needs and incorporate their opinions in forming laboratory policies.

範例 iii.

As his graduate school advisor, I can not think of a more qualified individual for implementing your organization's innovative technology developments.

範例 iv.

Susan is widely respected throughout our company, as I have heard often, as her group leader. While collaborating with her in various activities, I became aware of her exemplary communication and leadership skills. Her passion and resolve to pursue a management career also impressed me. Additionally, her refined coordinating skills and direct communication style have facilitated the smooth implementation of several events and company policies. Her strong leadership potential will definitely prove to be a valuable asset to your company.

範例 v.

Mr. Chang has been an associate researcher at our laboratory for five years.

During this period, he has been under my supervision.

C.被推薦人跟求職有關的個人特質

範例 i.

Mary is a highly motivated individual. In our product development group, she was responsible for performing various experimental procedures and analyzing the results. Her diligence in collecting and organizing materials within the laboratory played an important role in our product development efforts. She has the unique ability to identify exactly what is required for a particular research objective. She also quickly understands the limitations of conventional research. Moreover, she undertook numerous experiments, attempting to solve problems from various angles. Remaining confident despite occasional setbacks, she persevered during experimental work, ultimately yielding commercial success.
She is extremely well prepared for any assigned task. In our product development efforts, she constantly reviewed the latest technological developments and discussed her observations in detail with colleagues. During our weekly group meetings, which often involved of journals and case reports, she actively participated, offering carefully composed questions and responses to other group members. Furthermore, her analytical skills are exemplary. Although occasionally unfamiliar with a particular technology development at the outset, she would analyze the most pertinent information and then quickly identify the project goals and anticipated results.

範例 ii.

As a student, he worked diligently to develop his natural talents and displayed seemingly endless energy while under my instruction. I was particularly struck by his total commitment to the task at hand. His intelligence, industry and dedication will undoubtedly support his future employment. Armed with a passion for science, Mr. Li actively participated in several National Science Council-sponsored research projects. His maturity and diligence helped him to focus, as evidenced by his strong analytical skills and sound ability to formulate opinions after synthesizing available knowledge. Undoubtedly, these capabilities significantly contributed to his academic achievements, but will also ensure his future success.

His diligent attitude to studying never ceased to amaze me. For instance, whenever encountering a research bottleneck, he consistently delved into reading and investigating the source of the problem while consulting with me on how to solve it. Since graduation, he has continued to maintain contact with several researchers in the field, discussing issues relating to their clinical or research experience. In addition, his critical thinking skills are remarkable, as evidenced by his ability to synthesize pertinent reading, identify limitations of previous literature and then state the logical next step from a unique perspective.

範例 iii.

Matt is methodical and thorough in the task set before him, regardless of whether it is academic or professional. Graduate study equipped him with the required knowledge skills and fundamental professional expertise in industrial management. The graduate curriculum markedly differed from his undergraduate curriculum, offering many opportunities for him to strengthen his research fundamentals. For instance, the theoretical and practical concepts taught in the graduate curriculum increased his ability to solve problems logically and straightforwardly. Additionally, the theoretical knowledge and practical laboratory experience gained at graduate school were equally important in allowing him to foster his fundamental research skills. Given his deep interest in quality control, he is committed to pursuing a career in industrial management. In sum, graduate school equipped him with much knowledge and logical competence to address problems in the workplace effectively - an attribute that you will find attractive to your company.

I am also impressed with Matt's intuition when adapting to new environments. At graduate school, while learning how to adopt different perspectives in approaching a particular problem during undergraduate training, he acquired several statistical and analytical skills. Doing so involved learning how to analyze problems, identify potential solutions, and implement those solutions according to concepts taught in the classroom. The graduate school curricula equipped him not only with the academic fundamentals required for a management-related career, but also with the workplace skills to meet the rigorous challenges of the intensely competitive hi-tech sector. Securing employment in your company would definitely allow him to realize fully his career aspirations.

範例 iv.

Susan is truly an adaptable individual. For instance, her familiarity with implementing different strategies for various purposes allowed her to transfer to a new position in which she was responsible for simplifying administrative procedures and effectively managing personnel. Consequently, group morale was significantly raised owing to increased departmental efficiency. By constantly pursuing her research interests, she actively remained abreast of the latest technological developments and strived diligently to grasp their practical implementations. As your company has distinguished itself in providing high-quality technology products and services, I believe that you will find that Susan's professional experience and knowledge skills can easily blend into your product development team's innovative efforts. As a member of your organization's highly qualified staff, she would offer much research expertise in her area of specialization.

Susan's strong desire to improve her research capabilities constantly is reflected in her active participation in a collaborative project for our company, in which she analyzed customer data and then accumulated it in a novel database for statistical software purposes. Carefully analyzing the data revealed unique facts about the company's particular circumstances, which provided a valuable reference for administrators who had to make marketing-related decisions. In a similar development, Susan spearheaded our department's development of a novel administrative procedure for classifying and simplifying customers' financial information, processed by our company's accounting division. A database containing detailed customer information is accessed while ensuring the confidentiality of such information, thus reducing administrative costs and the number of personnel involved.

範例 v.

John is a resourceful individual. While serving as a research assistant when first entering our laboratory, he quickly learned how to coordinate different aspects of a research project, such as filling out weekly progress reports, managing financial affairs and organizing regularly held seminars and report contents. In addition to providing him with several opportunities to corroborate what he had learned from textbooks in the classroom, participation in several of our laboratory's research projects allowed him to extend his knowledge skills to fields outside his academic studies to provide innovative solutions. Moreover, his research often involved deriving complex models and modifying laboratory practices to meet a specific research requirement. He also attended several international conferences, which further widened his professional exposure. Moreover, intensive laboratory training definitely enhanced his ability to respond effectively to unforeseen bottlenecks in research.

In addition to his rich academic training and sound knowledge skills, John has many strong personality traits, as evidenced by his positive attitude towards challenges as he continuously strives for higher standards. He brings to your company a decade of experience in the chemical industry. After receiving a Bachelor's degree in Chemistry from National Chung Hsing University in 1993, he secured employment as a chemical engineer at Johnson Chemical Company. In this capacity, he acquired advanced knowledge skills by actively participating in many process development-related projects. He later joined Dupree Chemical Company in 1996, where he served as senior engineer responsible for process control. I believe that you will find such experience to be valuable to your highly qualified staff.

D.信函結尾

範例 i.

I have no qualms in recommending this highly motivated individual for employment in your organization. Her creativity and cooperative nature will be a great asset to any future product development effort in which she is involved. Your corporation's great working environment, combined with the impressive number and diversity of training courses to maintain the competence of its employees in the market place, would definitely ensure the ongoing development of Mary's professional skills while helping to improve the living standard of your customers. Feel free to contact me if I can provide you with further information.

範例 ii.

I do not hesitate in most highly recommending Jerry for employment in your organization. Do not hesitate to contact me if I can provide you with any further insight into this highly qualified individual.

範例 iii.

I am quite confident of Matt's ability to contribute significantly to any collaborative effort in which he is involved in your company. Please contact me if I can be of further assistance.

範例 iv.

I, therefore, have no hesitation in strongly recommending this individual to your corporate family. Your Human Resources Department is welcome to contact me for further insight into this highly qualified candidate.

範例 v.

I fully endorse John in his desire to work in your company. Please contact me if I can provide any further insight into Mr. Chang's ability.

◎Sample Employment Recommendation Letter / 求職推荐信函範例

Dear Mr. Ling,

簡介Given her ability to grasp abstract concepts quickly and willingness to accept others' constructive criticism to be more effective in an assigned task, I highly recommend Kelly Lin for employment in your organization. 推薦人的資格 Responsible for evaluating her work performance evaluations at our company over the past four years, I am in a unique position to assess her character and willingness to collaborate closely with other colleagues. She has the unique characteristic of examining a diverse array of topics and, then, searching for their possible relationship in the workplace to increase productivity. Moreover, she also easily adjusts to various organizational positions, reaffirming my conviction that flexibility is essential to strong interpersonal relations. As your company is renowned for its strong organizational culture and management structure, I firmly believe that employing such an innovative individual would be of greatly benefit.

被推薦人跟求職有關的個人特質 Kelly' s strong commitment to the marketing profession is demonstrated by a project that she spearheaded to identify effective demographic variables and forecast accurately growth trends of our company' s products and services. Given its effectiveness and accuracy, financial planners have already used this model in estimating consumer demand in the hi-tech sector. Kelly also initiated a similar project aimed at developing an efficient product control system, capable of monitoring the manufacturing capacity of our factory production line. The system subsequently developed by her team not only evaluates precisely bottlenecks in production, but also determines immediately the current output and efficiency of the production line. Her success in these projects reflects her adeptness in generating beneficial results that will contribute to corporate revenues, an attribute which I believe that your company is seeking.

信函結尾 I am highly confident of Kelly's ability to meet the rigorous work demands of your research team. As your company offers a competitive work environment for highly skilled professionals, I believe that she will be a welcome addition to your corporate family. Feel free to contact me if I can provide you with further information.

Yours truly,

Joe Wang

About the Author

Born on his father's birthday, Ted Knoy received a Bachelor of Arts in History at Franklin College of Indiana (Franklin, Indiana) and a Master's degree in Public Administration at American International College (Springfield, Massachusetts). He is currently a Ph.D. student in Education at the University of East Anglia (Norwich, England). Having conducted research and independent study in New Zealand, Ukraine, Scotland, South Africa, India, Nicaragua and Switzerland, he has lived in Taiwan since 1989 where he has been a permanent resident since 2000.

Having taught technical writing in the graduate school programs of National Chiao Tung University (Institute of Information Management, Institute of Communications Engineering, Institute of Technology Management, Department of Industrial Engineering Management, Department of Transportation Managment and, currently, in the College of Management) and National Tsing Hua University (Computer Science, Life Science, Electrical Engineering, Power Mechanical Engineering, Chemistry and Chemical Engineering Departments) since 1989, Ted also teaches in the Institute of Business Management at Yuan Pei University. He is also the English editor of several technical and medical journals and publications in Taiwan.

Ted is author of *The Chinese Technical Writers' Series*, which includes An English Style Approach for Chinese Technical Writers, English Oral Presentations for Chinese Technical Writers, A Correspondence Manual for Chinese Technical Writers, An Editing Workbook for Chinese Technical Writers and Advanced Copyediting Practice for Chinese Technical Writers. He is also author of *The Chinese Professional Writers' Series*, which includes Writing

Effective Study Plans, Writing Effective Work Proposals, Writing Effective Employment Application Statements, Writing Effective Career Statements, Effectively Communicating Online and Writing Effective Marketing Promotional Material.

Ted created and coordinates the Chinese On-line Writing Lab (OWL) at www.cc.nctu.edu.tw/~tedknoy and www.chineseowl.idv.tw

柯泰德先生自1989年即在國立交通大學及清華大學任教研究所科技英文寫作課程。目前柯先生是元培科技大學的講師。著有五本科技英文寫作系列及七本應用英文寫作系列叢書。同時他也是數本科技期刊的英文編輯。柯先生目前正在攻讀英國東安格利亞大學的教育博士學位。柯泰德線上英文論文編修訓練服務 www.cc.nctu.edu.tw/~tedknoy以及www.chineseowl.idv.tw

Acknowledgments

Thanks to the following individuals for contributing to this book:

特別感謝以下人員的貢獻：

國立交通大學工業管理學系

唐麗英（教授）　魏　源　林姍慧　金新恩　王有志　林麗甄　裴善康
張志偉　蔡志偉

元培科技大學 經營管理研究所

許碧芳（所長）　王貞穎　李仁智　陳彥谷　胡惠眞　陳碧俞　王連慶
蔡玟純　高青莉　賴姝惠　李雅玎　戴碧美　楊明雄　陳皇助　林宏隆
鍾玠融　李昭蓉　許美菁　葉伯彥　林羿君　吳政龍　鄭凱元　黃志斌
郭美萱　李尉誠　陳靜怡　盧筱嵐　鄭彥均　劉偉翔　彭廣興　林宗瑋
巫怡樺　朱建華

元培科科技大學 影像醫學研究所

王愛義（所長）　周美榮　顏映君　林孟聰　張雅玲　彭薇莉　張明偉
李玉綸　聶伊辛　黃勝賢
張格瑜　龔慧敏　林永健　呂忠祐　李仁忠　王國偉　李政翰　黃國明
蔡明輝　杜俊元　丁健益　方詩涵　余宗銘　劉力瑛　郭明杰

元培科技大學 生物技術研究所

陳媛孃（所長）　范齡文　彭姵華　鄭啓軒　許凱文　李昇憲　陳雪君
鄭凱暹　尤鼎元　陳玉梅　鄭美玲　郭軒中　朱芳儀　周佩穎　吳佳眞

國立交通大學管理學院

My technical writing students in the Department of Computer Science and Institute of Life Science at National Tsing Hua University, as well as the College of Management at National Chiao Tung University are also appreciated. Thanks also to Seamus Harris and Bill Hodgson for reviewing this book.

科技英文寫作系列之一

精通科技論文（報告）寫作之捷徑

An English Style Approach for Chinese Technical Writers （修訂版）

作者：柯泰德（Ted Knoy）

內容簡介

使用直接而流利的英文會話
讓您所寫的英文科技論文很容易被了解
提供不同形式的句型供您參考利用
比較中英句子結構之異同
利用介系詞片語將二個句子連接在一起

萬其超／李國鼎科技發展基金會秘書長

本書是多年實務經驗和專注力之結晶，因此是一本坊間少見而極具實用價值的書。

陳文華／國立清華大學工學院院長

中國人使用英文寫作時，語法上常會犯錯，本書提供了很好的實例示範，對於科技論文寫作有相當參考價值。

徐　章／工業技術研究院量測中心主任

這是一個讓初學英文寫作的人，能夠先由不犯寫作的錯誤開始再根據書中的步驟逐步學習提升寫作能力的好工具，此書的內容及解說方式使讀者也可以無師自通，藉由自修的方式學習進步，但是更重要的是它雖然是一本好書，當您學會了書中的許多技巧，如果您還想要更進步，那麼基本原則還是要常常練習，才能發揮書的精髓。

Kathleen Ford, English Editor, Proceedings(Life Science Divison), National Science Council

The Chinese Technical Writers Series is valuable for anyone involved with creating scientific documentation.

※若有任何英文文件修改問題，請直接與柯泰德先生聯絡：（03）5724895

科技英文寫作系列之二

作好英語會議簡報

English Oral Presentations for Chinese Technical Writers

作者：柯泰德（Ted Kony）

內容簡介

本書共分十二個單元，涵括產品開發、組織、部門、科技、及產業的介紹、科技背景、公司訪問、研究能力及論文之發表等，每一單元提供不同型態的科技口頭簡報範例，以進行英文口頭簡報的寫作及表達練習，是一本非常實用的著作。

李鍾熙／工業技術研究院化學工業研究所所長

一個成功的科技簡報，就是使演講流暢，用簡單直接的方法、清楚表達內容。本書提供一個創新的方法（途徑），給組織每一成員做爲借鏡，得以自行準備口頭簡報。利用本書這套有系統的方法加以練習，將必然使您信心備增，簡報更加順利成功。

薛敬和／IUPAC台北國際高分子研討會執行長
國立清華大學教授

本書以個案方式介紹各英文會議簡報之執行方式，深入簡出，爲邁入實用狀況的最佳參考書籍。

沙晉康／清華大學化學研究所所長
第十五屆國際雜環化學會議主席

本書介紹英文簡報的格式，值得國人參考。今天在學術或工商界與外國接觸來往均日益增多，我們應加強表達的技巧，尤其是英文的簡報應具有很高的專業水準。本書做爲一個很好的範例。

張俊彥／國立交通大學電機資訊學院教授兼院長

針對中國學生協助他們寫好英文的國際論文參加國際會議如何以英語演講、內容切中要害特別推薦。

※若有任何英文文件修改問題，請直接與柯泰德先生聯絡：（03）5724895

科技英文寫作系列之三

英文信函參考手冊

A Correspondence Manual for Chinese Technical Writers

作者：柯泰德（Ted Knoy）

內容簡介

本書期望成為從事專業管理與科技之中國人，在國際場合上溝通交流時之參考指導書籍。本書所提供的書信範例（附磁碟片），可為您撰述信件時的參考範本。更實際的是，本書如同一「寫作計畫小組」，能因應特定場合（狀況）撰寫出所需要的信函。

李國鼎／總統府資政

我國科技人員在國際場合溝通表達之機會急遽增加，希望大家都來重視英文說寫之能力。

羅明哲／國立中興大學教務長

一份表達精準且適切的英文信函，在國際間的往來交流上，重要性不亞於研究成果的報告發表。本書介紹各類英文技術信函的特徵及寫作指引，所附範例中肯實用，爲優良的學習及參考書籍。

廖俊臣／國立清華大學理學院院長

本書提供許多有關工業技術合作、技術轉移、工業資訊、人員訓練及互訪等接洽信函的例句和範例，頗爲實用，極具參考價值。

于樹偉／工業安全衛生技術發展中心主任

國際間往來日益頻繁，以英文有效地溝通交流，是現今從事科技研究人員所需具備的重要技能。本書在寫作風格、文法結構與取材等方面，提供極佳的寫作參考與指引，所列舉的範例，皆經過作者細心的修訂與潤飾，必能切合讀者的實際需要。

※若有任何英文文件修改問題，請直接與柯泰德先生聯絡：（03）5724895

科技英文寫作系列之五

科技英文編修訓練手冊【進階篇】

Advanced Copyediting Practice for Chinese Technical Writers

作者：柯泰德（Ted Knoy）

內容簡介

本書延續科技英文寫作系列之四「科技英文編修訓練手冊」之寫作指導原則，更進一步把重點放在如何讓作者想表達的意思更明顯，即明白寫作。把文章中曖昧不清全部去除，使閱讀您文章的讀者很容易的理解您作品的精髓。

本手冊同時國立清華大學資訊工程學系非同步遠距教學科技英文寫作課程指導範本。

張俊彥 / 國立交通大學校長暨中研院院士

對於國內理工學生及從事科技研究之人士而言，可說是一本相當有用的書籍，特向讀者推薦。

蔡仁堅 / 前新竹市長

科技不分國界，隨著進入公元兩千年的資訊時代，使用國際語言撰寫學術報告已是時勢所趨；今欣見柯泰德先生致力於編撰此著作，並彙集了許多實例詳加解說，相信對於科技英文的撰寫有著莫大的裨益，特予以推薦。

史欽泰 / 工研院院長

本書即以實用範例，針對國人寫作的缺點提供簡單、明白的寫作原則，非常適合科技研發人員使用。

張智星 / 國立清華大學資訊工程學系副教授、計算中心組長

本書是特別針對系上所開科技英文寫作非同步遠距教學而設計，範圍內容豐富，所列練習也非常實用，學生可以配合課程來使用，在時間上更有彈性的針對自己情況來練習，很有助益。

劉世東 / 長庚大學醫學院微生物免疫科主任

書中的例子及習題對閱讀者會有很大的助益。這是一本研究生必讀的書，也是一般研究者重要的參考書。

※若有任何英文文件修改問題，請直接與柯泰德先生聯絡：（03）5724895

有效撰寫英文讀書計畫
Writing Effective Study Plans

作者：柯泰德（Ted Knoy）

內容簡介

本書指導準備出國進修的學生撰寫精簡切要的英文讀書計畫，內容包括：表達學習的領域及興趣、展現所具備之專業領域知識、敘述學歷背景及成就等。本書的每個單元皆提供視覺化的具體情境及相關寫作訓練，讓讀者進行實際的訊息運用練習。此外，書中的編修訓練並可加強「精確寫作」及「明白寫作」的技巧。本書適用於個人自修以及團體授課，能確實引導讀者寫出精簡而有效的英文讀書計畫。

本手冊同時為國立清華大學資訊工程學系非同步遠距教學科技英文寫作課程指導範本。

于樹偉／工業技術研究院主任

《有效撰寫讀書計畫》一書主旨在提供國人精深學習前的準備，包括：讀書計畫及推薦信函的建構、完成。藉由本書中視覺化訊息的互動及練習，國人可以更明確的掌握全篇的意涵，及更完整的表達心中的意念。這也是本書異於坊間同類書籍只著重在片斷記憶，不求理解最大之處。

王　玫／工業研究技術院、化學工業研究所組長

《有效撰寫讀書計畫》主要是針對想要進階學習的讀者，由基本的自我學習經驗描述延伸至未來目標的設定，更進一步強調推薦信函的撰寫，藉由圖片式訊息互動，讓讀者主動聯想及運用寫作知識及技巧，避免一味的記憶零星的範例；如此一來，讀者可以更清楚表明個別的特質及快速掌握重點。

應用英文寫作系列之二

有效撰寫英文工作提案
Writing Effective Work Proposals

作者：柯泰德（Ted Knoy）

內容簡介
許多國人都是在工作方案完成時才開始撰寫相關英文提案，他們視撰寫提案為行政工作的一環，只是消極記錄已完成的事項，而不是積極的規劃掌控未來及現在正進行的工作。如果國人可以在撰寫英文提案時，事先仔細明辨工作計畫提案的背景及目標，不僅可以確保寫作進度、寫作結構的完整性，更可兼顧提案相關讀者的興趣強調。本書中詳細的步驟可指導工作提案寫作者達成此一目標。 書中的每個單元呈現三個視覺化的情境，提供國人英文工作提案寫作實質訊息，而相關附加的寫作練習讓讀者做實際的訊息運用。此外，本書也非常適合在課堂上使用，教師可以先描述單元情境而讓學生藉由書中練習循序完成具有良好架構的工作提案。書中內容包括：1.工作提案計畫（第一部分）：背景 2.工作提案計畫（第二部分）：行動 3.問題描述 4.假設描述 5.摘要撰寫（第一部分）： 簡介背景、目標及方法 6.摘要撰寫（第二部分）： 歸納希望的結果及其對特定領域的貢獻 7.綜合上述寫成精確工作提案。

唐傳義／國立清華大學資訊工程學系主任

本書重點放在如何在工作計畫一開始時便可以用英文來規劃整個工作提案，由工作提案的背景、行動、方法及預期的結果漸次教導國人如何寫出具有良好結構的英文工作提案。如此用英文明確界定工作提案的程序及工作目標更可以確保英文工作提案及工作計畫的即時完成。對工作效率而言也有助益。

在國人積極加入WTO之後的調整期，優良的英文工作提案寫作能力絕對是一項競爭力快速加分的工具。

※若有任何英文文件修改問題，請直接與柯泰德先生聯絡：（03）5724895

應用英文寫作系列 07

管理英文

作　　者 / 柯泰德（Ted Knoy）
出 版 者 / 揚智文化事業股份有限公司
發 行 人 / 葉忠賢
總 編 輯 / 閻富萍
執行編輯 / 胡琡珮
地　　址 / 台北縣深坑鄉北深路三段 260 號 8 樓
電　　話 / (02)86626826
傳　　真 / (02)2664-7633
E-mail / service@ycrc.com.tw
印　　刷 / 鼎易印刷事業股份有限公司
ISBN / 978-957-818-898-3
初版一刷 / 2008 年 10 月
定　　價 / 新台幣 450 元

國家圖書館出版品預行編目資料

管理英文＝Effective management communication
/ 柯泰德（Ted Knoy）作. -- 初版. -- 臺北縣
深坑鄉：揚智文化, 2008.10
面；　公分. --（應用英文寫作系列；7）

ISBN 978-957-818-898-3（平裝）

1.商業英文 2.商業應用文

493.6　　97020777